howstuffworks.com

MW00366512

The Real Science of SEX APPEAL

Why We Love, Lust, and Long for Each Other

sourcebooks

Published by Sourcebooks, Inc.
P.O. Box 4410, Naperville, Illinois 60567-4410
(630) 961-3900
Fax: (630) 961-2168
www.sourcebooks.com

Library of Congress Cataloging-in-Publication Data
The real science of sex appeal : why we love, lust, and long for each other / HowStuffWorks.
     pages cm
1. Sex (Psychology) 2. Sex (Biology) 3. Sex. 4. Mate selection. I. HowStuffWorks, Inc.
 BF692.R43 2014
 155.3'1--dc23

2014030134

Printed and bound in the United States of America.
VP 10 9 8 7 6 5 4 3 2 1

Also by *HowStuffWorks*

*Stuff You Missed in History Class: A Guide to
History's Biggest Myths, Mysteries, and Marvels*

*Future Tech, Right Now: X-Ray Vision, Mind
Control, and Other Amazing Stuff from Tomorrow*

*The Science of Superheroes and Space Warriors:
Lightsabers, Batmobiles, Kryptonite, and More!*

CONTENTS

INTRODUCTION: WHY DO WE LOVE?

L ove—is it really all we need? Scientists and song-writers agree that love is one of the most important aspects of being human, but why do we get that familiar head rush when we meet someone we like? What is love anyway? Is it just an emotion…or something greater? In this book, we take a look at the crazy, sexy chemistry and science behind why we lust and long for each other, and fall in love (and sadly, sometimes out of it), and what the future holds for sex, love, and relationships. First, let's start with the most basic question: why do we love?

As unromantic as it may seem, there's a solid scientific reason behind our capacity to love. Love is actually a moti-vating goal for humans, and our behavior can be explained by our attempts to achieve this goal. The human motivation system is linked to the reward system in the brain. Once we achieve a goal, the brain releases dopamine into a region of the reward system called the nucleus acumens. We experience this as a profound sense of pleasure and excitement—the types of sensations one associates with the experience of romantic love.

In 2005, a groundbreaking academic study using functional magnetic resonance imaging (fMRI) found visual evidence that supports the view of love as motivation. The researchers found that when seventeen young participants were shown a photo of the person they loved, regions of the brain responsible for motivating and rewarding began to function. In other words, romantic love motivates people, and the motivation toward this goal—loving and being loved—is fueled by the brain's reward system.

The imaging also showed that while the emotional centers of the brain were active, no distinct pattern of emotions was followed. This finding counters the long-standing view that love is based in emotion. Instead, love seems to spring from our goal-seeking behavior, and the emotions that we attach to it come second to our motivation.

This is a nice, tidy biological explanation for love, but how did our strong capacity to love evolve, and why?

Evolution and Love

The question of why we're equipped to love has already been answered via evolutionary theory: we love because we're meant to reproduce. Species continue through reproduction, and continuation of the species is paramount in evolution. Since mating is the ultimate goal, feelings of romantic love are merely a vehicle toward this goal. Yet, the

2005 study found that the areas that cause sexual arousal in the brain aren't fully active as people fall in love. The two regions overlap, but the experiences aren't the same.

This doesn't disprove the idea that love exists to foster reproduction, but it certainly raises new questions. Specifically, why do we continue to feel love even after we've reproduced? Well, the current answer is also based in evolution. The combination of reward and attachment actually leads to a lasting *addiction* for a particular individual— our partner.

Because of the association with reward motivation and its attendant releases of dopamine (a brain chemical we'll explore in more detail later), that initial rush of romantic love resembles addiction rather than emotion. Over time, however, other neurotransmitters may play a larger role in forming long-term attachment that lasts beyond our reproductive years.

The chemicals vasopressin and oxytocin help humans and about 3 percent of other mammal species to experience lasting, monogamous love. These two chemicals are associated with our ability to form memories of others and help us recognize other people. They're also released, along with dopamine, during sex.

This combination of dopamine (which induces feelings of pleasure), oxytocin (which is associated with feelings of attachment), and vasopressin (which also promotes attachment and also allows social recognition) leads to a learned behavior by which we actually become addicted to our mate. Regardless of whether it's the sight of the person we're in love with or the injection of some drug, if both trigger similar releases, humans can experience both similarly and become addicted as well.

These same chemicals may also play a role in familial love, like that between a parent and child or among siblings. The chemical oxytocin, for example, plays a role in parental bonding. It's released in mothers during childbirth, and it plays a role in the production and release of breast milk.

We experience love, then, to foster the relationships that may lead to reproduction and to maintain relationships with the offspring borne from those relationships. Seems like a fairly straightforward explanation, no? Well, it turns out that a lot of sexy science has to happen for us to love. Let's start by breaking down the crazy chemistry behind some of the coolest parts of desire.

1

LET'S GET PHYSICAL

'HAVE WE MET BEFORE?': HOW FLIRTING WORKS

I magine that you have no idea what flirting is. If you haven't flirted yourself or seen it happen (either in real life, in the movies, or on TV), you might wonder exactly what those two people are doing. They're showing interest in each other, but they don't actually come out and say it. In fact, it's usually considered crass and crude to do so. Instead, they dance around the issue—joking, complimenting each other, and using physical cues to show their true intentions.

At its most basic, flirting is simply another way that two people can interact closely with each other. But when you get into the intentions behind flirting and exactly what flirting entails, things get *much* more interesting. Flirting doesn't have to be romantic or sexual. Sometimes it's just friendly banter without any other intention. Sometimes one person has romantic intentions and the other one only has sexual ones—or doesn't even realize that he's being flirted with.

It can be difficult to know when someone is flirting with you or who might be receptive to your flirting.

Misunderstanding the signals can lead to some uncomfortable and embarrassing situations. After all, the most important aspect of flirting is the intention behind it. Sometimes the words used are very innocent, but the speaker's delivery, expression, or mannerisms make them appear flirtatious. What's a guy or gal to do?

Let's start by examining the standard signs of flirting. We'll also look at the biological factors that lead to flirting and explore how flirting has changed through the years. By the end, you'll know whether those batting eyelashes are saying, "Come hither" or "Back off, buddy."

Flirting Signs and Signals

How do you know when someone is flirting with you? Although it's different for everybody, there are some common signs. Here are a few clues:

- ♥ Using your name a lot in conversation.
- ♥ Complimenting you.
- ♥ Asking about your interests.
- ♥ Touching your arm or knee.
- ♥ Leaning in while talking.
- ♥ Standing closely.
- ♥ Smiling a lot.

Just one of these actions, or even a few of them together, would probably not constitute flirting. But if someone compliments you, smiles often, leans in closely, and brushes your arm as he talks to you, there's a good possibility that he's flirting.

To compound the issue, there are differences between how women and men flirt. For example, some women bat their eyelashes or run their fingers through their hair. Men are more likely to make bold, aggressive gestures, like intense eye contact. In addition, they are more likely to flirt out of sexual interest, while women often flirt to test men's intentions, using ambiguous gestures.

Cultural anthropologist Kate Fox of the Social Issues Research Centre has coined a term for these ambiguous flirting gestures, like hair touching. They're called *protean signals*, named for the Greek shape-shifting god Proteus. If a woman uses these gestures and learns that the man isn't interested, then she can always play them off as not being flirtatious.

If you want to flirt, you could try any of these methods. But you have to watch carefully for the other person's reaction. If he leans away when you lean forward, or if he doesn't engage in conversation despite your best attempts, then he might not be interested. On the other hand, he could simply be shy and taken aback by your interest. Or he could think that you're just being friendly when you're actually interested in more than just being friends. How do you know for sure? There are no definite rules when it comes to flirting, because every situation is different. For better or worse, it's as complex and tricky as dating in general.

So although there are some obvious signs of flirting, it can still be a very messy endeavor. In a 2006 article for the *Daily Mail*, reporter Danielle Gusmaroli wrote about trying a method employed by a successful flirter that she'd interviewed:

On leaving the bar, I spot a road cleaner across the street and smile warmly. He smiles back and I hold his gaze for an agonizing four seconds, look away, and (cringe) look back. He smiles appreciatively and I scuttle off trying not to laugh.

To my horror, he pegs it across the road to my side. With a penetrating stare he asks: "Sorry, do I know you?"

I apologize, trying to back off. "Sorry, my mistake. I thought you were someone else."

"Give me your number," he demands. I decline. He becomes angry. "You were coming on to me, weren't you?" I panic and run.

What happened? Gusmaroli was trying to flirt, but she wasn't really interested in the road cleaner. It took him a while to recognize the flirting, and when he did, he seemed to feel like she "owed him" her phone number.

In the next section, we'll look at the science of

flirting—what's happening in your brain and body when you flirt and how flirting works in other species.

FLIRTING VS. SEXUAL HARASSMENT

Some people confuse sexual harassment with flirting, and vice versa. The most important factor here is perception. It's sexual harassment when it leaves the person feeling offended, demeaned, or violated. The bottom line is that if the object of someone's flirtation makes it very clear that the flirting is unwelcome and unwanted, but the person continues to flirt, then he or she is guilty of harassment.

The Science of Flirting

There's a lot going on under the surface when we flirt. Yes, we're sending the message that we're interested, but why do those specific gestures say "I'm interested in you," and what do they really say about us? According to scientists, it all comes down to our inherent desire to reproduce, as previously mentioned. When we flirt, we're giving off information about how fit we are to procreate, as well as the state of our overall health. There are also specific aspects of our appearance that make us more attractive to others.

Some of the "female" signs of flirting, such as angling her body and sticking out her hips, are attempts to draw attention to her pelvis and its suitability for carrying a child. In addition, men tend to be more attracted to women with a certain hip-to-waist ratio (specifically, the waist must be no more than 60 to 80 percent of the hip circumference). This is also an indication of fertility.

When a man makes intense eye contact and smiles often, he is attempting to show that he is both virile and dependable. Women are attracted to prominent, square jaws, which are indicative of a man's power and strength. Scientists point out that features like square jaws in human males have a connection to prominent features in the animal kingdom. Male peacocks attract females with their elaborate plumage. Male cardinals are bright red, and stags have large horns. Because these features require additional biological resources and also tend to make these animals more visible to their predators, an impressive display shows that these animals are strong.

When we're flirting with someone who fits the bill for us, the limbic system takes over (the same system responsible for our fight-or-flight response). We operate on emotion and instinct. If we only governed flirting with the most rational part of our brains, we might not ever flirt—or even get a date. In fact, according to Antonio Damasio, MD,

PhD, there's a connection between brain damage and flirting. He states that "people with damage to the connection between their limbic structures and the higher brain are smart and rational—but unable to make decisions."

Still, we're not just animalistic in our flirting behavior. The ability to carry on a conversation and engage in the joking back-and-forth of flirting also indicates our intelligence, which is always attractive. In the next section, we'll look at how flirting has changed over the years and how technology has led to new ways of flirting.

FLIRTING ETIQUETTE

Although there are some basic flirting gestures, the timing may be different in other countries. For example, in Germany flirting is much more subtle on the part of both women and men. German women find American men far too forward. German men often misunderstand the intentions of American women—they think they're being flirted with when they're not.

Flirting through the Ages

Today, we—meaning, most Western societies—aren't really shocked by typical flirtatious behaviors. But if a person from the Victorian era witnessed the knee-touching, lip-licking, and winking that goes on today, he or she might be extremely scandalized. Flirting has gone from very prescribed sets of behaviors to off-the-cuff text and Facebook messages.

Some elements of flirting are eternal. A recent translation of the *Kama Sutra*, the classic Hindu sexual instruction manual, includes several ways for men to flirt successfully with women. One includes the suggestion that when a man and woman are "playing in the water, he dives underwater at some distance from her, comes up close to her, touches her, and dives underwater again." This sounds like typical flirty behavior between kids.

An 1881 book advising young Victorian men and women on manners and etiquette includes several guidelines for courtship. First and foremost, a "gentleman should not be introduced to a lady, unless her permission has been previously obtained." Once he is introduced, he has the freedom to call on her and accompany her to "concerts, operas, balls, etc." However, a "gentleman who does not contemplate matrimony should not pay too exclusive attention to any one lady."

A proper Victorian woman "will not too eagerly receive the attentions of a gentleman, no matter how much she admires him; nor, on the other hand, will she be so reserved as to altogether discourage him." The book goes on to describe how men should only take the hand of a woman when she offers it, and that the kiss, "the most affectionate form of salutation, is only proper among near relations and dear friends."

Today's teenagers surely would not bother with something as seemingly trivial as shaking hands to send the correct messages. Over the years, flirting has evolved from careful, measured gestures to cutting to the chase. It's completely common for a fifteen-year-old boy to text a line like "how far have u gone" to a girl after a few days of flirtatious texting, according to a 2006 article in *Time* magazine. Sometimes the girl replies honestly, or she might say, "how far have u gone. ill tell u if u tell me." This could lead to meeting up and making out.

Face-to-face meetings often result in exchanges of e-mail addresses, or Facebook info instead of phone numbers. It's also not uncommon for strangers to meet online and exchange flirtatious banter. But online flirters beware: in December 2007, a manufacturer of antivirus software discovered a Russian virus that invaded chat rooms. Once in, the virus chatted with users and flirted with them so convincingly that some women shared their photos and phone numbers.

WHAT'S MY LINE?

There really is no such thing as a good pickup line. But definitely don't bother trying any of these—they are some of the oldest in the book:

+ I seem to have lost my phone number. Can I have yours?
+ If I could arrange the alphabet, I would put U and I together.
+ Is it hot in here, or is it just you?
+ And the classics: "What's your sign?" and "Do you come here often?"

PHEROMONES AT FIRST SIGHT? THE TOP FIVE PHYSICAL SIGNS OF ATTRACTION

E ven if you can't describe your ideal romantic partner, you'll know him or her on sight. According to Rutgers University anthropologist Helen Fisher, the human body is such a finely tuned attraction-seeking machine that it takes only one second to intuitively decide whether someone's physically hot or not. Upon closer inspection, we might change our minds, or we just might have found what we've been looking for all along. Your body will then produce physical signs that grab your attention and direct it toward the crush in question. Which lovesick symptoms should you look out for to know if you've found a catch?

1. Be Still, My Beating Heart

Why do literature and art always associate romance and the heart? Because our hearts are set aflutter, pulses literally racing, at the sight of someone attractive. In fact, the heart-attraction relationship is so potent, studies have found, that increasing someone's heart rate and then putting him or her near a good-looking stranger can artificially ignite a flame of affection.

Per usual, the brain is ultimately responsible for this physiological response, not Cupid and his archery acumen. During early-stage romantic love—scientific terminology for the honeymoon phase—the brain releases norepinephrine whenever we're around a love interest to shake us into action. That adrenaline-like neurotransmitter spurs our motivational decision-making, possibly prodding us to chat up Mr. or Ms. Right. Meanwhile, our adrenaline-addled hearts are likely pumping faster than usual to get us through the taxing ordeal.

2. Sweating the Small Stuff

If you're introduced to someone who immediately makes your heart go gaga, it might be best to avoid a handshake. Sweating palms are a classic physiological response to attraction. The same cocktail of chemicals that prods our pulses also stokes our sweat glands. Dopamine, norepinephrine, and serotonin—collectively known as monoamines—combine to produce feelings of excitement, with a side of breathlessness and moist hands. Norepinephrine, in particular, is the culprit for goading our sweat glands into activation, and since our palms are riddled with up to three thousand miniscule sweat glands per square inch, they can quickly become a telltale signal of sexual interest.

Men may be stricken with sweaty palms more often than women. Anthropologist Helen Fisher suggests that since men are more visually stimulated than women, their brains dole out bigger doses of monoamines.

3. Be Mine, Baritone

Repeated studies have confirmed that heterosexual women prefer deeper voices whispering sweet nothings in their ears. In addition to associating lower-pitched male voices with masculinity, women associate those bass notes with author-ity, larger body size, and physical attractiveness. Fortunately for the tenors out there, an Australian study published in December 2011 at least debunked the notion that a deeper voice intimated superior sperm quality.

Perhaps since deeper-pitched voices have attracted such a sexy reputation, people may lower their registers when speaking to their special someones. In a 2010 study, male and female study participants were asked to record messages to be played for fictional recipients. Researchers showed individual participants photos of fictional message recipients and found that the more attractive participants rated the fictional recipients, the more likely they were to deepen their voices. But a conflicting study found that the more tantalizing the male face, the higher—not lower—women raised their vocal pitch. Either way, it seems we attempt to fine-tune our voices to sound like sweet music to our beloveds' ears.

> 4. Jeepers Peepers

Dusty Springfield wasn't talking nonsense when she sang about "The Look of Love." The chart-topping blond with the golden voice belted out scientifically plausible lyrics about come-hither eyes, as studies have shown that our pupils play an active role in signaling attraction. When we spot a comely face, our brains release dopamine, which triggers pupil dilation. Thanks to the surge of dopamine in our brains that excites the nerve endings in our eyes, the pupil muscles contract and dilate our peepers.

But pupil preferences aren't uniform across the board, and bigger isn't always better. While heterosexual men find women with larger dilated pupils more feminine and beautiful, most straight women opt for medium male dilations that signal sexual interest, but not to a potentially violent extreme. However, women who tend to engage in short-term sexual relationships with "bad boys" were googly-eyed for the larger pupils as well.

> 5. Copycats

Nineteenth-century British writer and aphorism documentarian Charles Caleb Colton ushered the phrase "Imitation is the sincerest form of flattery" into common usage in 1820. Come to find out, Colton's adage applies quite well to interpersonal attraction. When people interact in dating scenarios, and things are going well, mirroring of body language often happens subconsciously. For instance, when someone leans in close to the dinner table, the other person follows suit.

Better yet, without us knowing it, these subtle gestures serve to stoke each other's romantic egos. A 2009 study on mimicry in a speed-dating environment revealed that men gave more favorable ratings to women who slightly mirrored their verbal and nonverbal patterns. Scratching their faces after the men scratched their faces, for instance, ultimately increased the women's sexual attractiveness after the five-minute interaction.

If that body language exchange sparks a long-lasting relationship, men's and women's bodies tend to play copycat as they age together, as well. According to a 2006 study, the longer couples stick together, well after the jittery symptoms of attraction have calmed, the more they physically begin to look alike.

PUCKER UP: HOW KISSING WORKS

When you really think about it, kissing is pretty gross. It involves saliva and mucous membranes, and it may have historical roots in chewed-up food! Experts estimate that hundreds or even millions of bacterial colonies move from one mouth to another during a kiss. Doctors have also linked kissing to the spread of diseases like meningitis, herpes, and mononucleosis. Gross, right?

Yet anthropologists report that 90 percent of people in the world kiss. Most people look forward to their first romantic kiss and remember it for the rest of their lives. Parents kiss children, worshippers kiss religious artifacts, and couples kiss each other. Some people even kiss the ground when they get off an airplane.

So how does one gesture come to signify affection, celebration, grief, comfort, and respect all over the world? No one knows for sure, but anthropologists think kissing might have originated with human mothers feeding their babies much the way birds do. Mothers would chew the food and then pass it from their mouths to their babies' mouths. After the babies learned to eat solid food, their mothers may have kissed them to comfort them or to show affection.

In this scenario, kissing is a learned behavior, passed from generation to generation. We do it because we learned how from our parents and from the society around us. There's a problem with this theory, though. Women in a few modern indigenous cultures feed their babies by passing chewed food mouth-to-mouth. But in some of these cultures, no one kissed each other for affection until Westerners introduced the practice.

Other researchers believe instead that kissing is instinctive. They use bonobo apes, which are closely related to humans, to support this idea. Bonobos kiss one another frequently. Regardless of sex or status within their social groups, bonobos kiss to reduce tension after disputes, to reassure one another, to develop social bonds, and sometimes for no clear reason at all. Some researchers believe the kissing primates prove that the desire to kiss is instinctive.

Several other animal species have behaviors that resemble kissing. Many mammals lick one another's faces. Birds touch one another's bills, and snails caress one another's antennae. In some cases, the animals are grooming one another rather than kissing. In others, they're smelling scent glands that are located on faces or in mouths. Regardless, when animals touch each other in this way, they're often showing signs of trust and affection or developing social bonds.

KISS AND TELL?

Modern research suggests that just about every culture on the planet kisses. However, anthropologists and ethnologists have described a few cultures in Asia, Africa, and South America that do not kiss at all. Some of these cultures view kissing as disgusting or distasteful. However, other researchers point out that these societies may view kissing as too private to discuss with strangers. In other words, they may kiss but not talk about it.

The Effects of Kissing

While researchers aren't exactly sure how or why people started kissing, they do know that romantic kissing affects most people profoundly. The Kinsey Institute describes a person's response to kissing as a combination of three factors:

♥ Your psychological response depends on *your mental and emotional state*, as well as *how you feel about the person who is kissing you*. Psychologically, kissing someone you want to kiss will generally encourage feelings of attachment and affection. If you're kissing someone you don't like, or you're kissed against your will, your psychological response will be completely different.

♥ *The culture in which you grew up* plays a big part in how you feel about kissing. In most Western societies, people are conditioned to look forward to and enjoy kissing. The behavior of the people around you, depictions in the media, and other social factors can dramatically affect how you respond to being kissed.

These factors play a part in all kisses, not just those that are romantic in nature. In other words, when a mother kisses her child's bruise to make it feel better, psychological, physical, and social factors play a part in both people's reactions. The same is true when friends kiss as a greeting, worshippers kiss religious symbols, or siblings kiss and make up after an argument. Even though some kisses are platonic and others are romantic, they generally have one thing in common—they are inspired by and tend to inspire feelings we think of as positive.

Regardless of exactly how people got the idea to kiss or what they mean when they do it, anthropologists are fairly sure that people started kissing thousands of years ago. We'll look at the history of kissing in more detail next.

WHAT AN INCREDIBLE SMELL!

People in some cultures rub one another's noses or cheeks rather than, or in addition to, kissing. Anthropologists theorize that this "Eskimo kiss" grew from people smelling one another's faces much the way animals do.

Smooching throughout History: The History of the Kiss

Historians really don't know much about the early history of kissing. Four Vedic Sanskrit texts, written in India around 1500 BC, appear to describe people kissing. This doesn't mean that nobody kissed before then, and it doesn't mean that Indians were the first to kiss. Artists and writers may just have considered kissing too private to depict in art or literature.

After its first mention in writing, kissing didn't appear much in art or literature for a few hundred years. The Indian epic poem *Mahabharata*—which was passed down orally for several hundred years before being written down and standardized around 350 AD—describes kissing on the lips as a sign of affection. The Indian erotic text, the *Kama Sutra*, written in the sixth century AD, also describes a variety of kisses. Anthropologists who believe that kissing is a learned behavior theorize that the Greeks learned about it when Alexander the Great invaded India in 326 BC.

There aren't many records of kissing in the Western world until the days of the Roman Empire. Romans used kisses to greet friends and family members. Citizens kissed their rulers' hands. And, naturally, people kissed their romantic partners. The Romans even came up with three different categories for kissing:

- *Osculum* was a kiss on the cheek.
- *Basium* was a kiss on the lips.
- *Savolium* was a deep kiss.

The Romans also started several kissing traditions that have lasted to the present day. In ancient Rome, couples became betrothed by kissing passionately in front of a group of people. This is probably one reason why modern couples kiss at the end of wedding ceremonies. Additionally, although most people today think of love letters as "sealed with a kiss," kisses were used to seal legal and business agreements. Ancient Romans also used kissing as part of political campaigns. However, several "kisses for votes" scandals in eighteenth-century England led—in theory—to candidates kissing only the very young and very old.

KISSING UNDER THE MISTLETOE

Today, some people seem to spend the holiday season waiting under the mistletoe in hope of kissing whoever passes by. But until the 1400s, kissing under mistletoe was a big commitment. Such kisses often meant that a couple was engaged.

Kissing also played a role in the early Christian Church. Christians often greeted one another with an *osculum pacis*, or kiss of peace. According to this tradition, the kiss of peace caused a transfer of spirit between the two people kissing. Most researchers believe the purpose of this kiss was to establish familial bonds between the members of the church and to strengthen the community.

Until 1528, the kiss of peace was part of Catholic mass. In the thirteenth century, the Catholic Church substituted a pax board, which the congregation kissed instead of kissing one another. The Protestant Reformation in the 1500s removed the kiss from Protestant services entirely. The kiss of peace doesn't typically play a role in modern Christian religious services, although some Christians do kiss religious symbols, including the Pope's ring.

But not all kisses have been happy events. Works of

literature like *Romeo and Juliet* have portrayed kisses as dangerous or deadly when shared between the wrong people. Some folklorists and literary critics view vampirism as symbolic of the physical and emotional dangers that can come from kissing the wrong person.

Also, while most cultures around the world kiss today, many have different views about when and where kissing is appropriate. In the 1990s, several news articles reported a trend among young people of kissing in public in Japan, where kissing had traditionally been viewed as a private activity.

KISS OF JUDAS

One of the Western world's most famous kisses is the kiss Judas Iscariot used to betray Jesus shortly before his crucifixion. This kiss had an influence on Christian spiritual practices. Early church sects omitted the kiss of peace—or abstained from kissing entirely—on Maundy Thursday. Maundy Thursday is the Thursday before Easter and the day used to commemorate the Last Supper, after which Judas betrayed Jesus in the Garden of Gethsemane.

⸖ Anatomy of a Kiss

Most people think about what to do when kissing another person, but few ponder all the technical details behind it. No matter who you're kissing or why, the basic kiss relies heavily on one muscle—the orbicularis oris, which runs around the outside of your mouth. Your orbicularis oris changes the shape of your mouth while you talk, and it puckers your lips when you kiss.

But the orbicularis oris is really just the tip of the iceberg. About two-thirds of people tip their heads to the right while kissing. Scientists believe this preference starts before we're born, when we tip our heads to the right in the womb. Muscles in your head, neck, and shoulders tilt your head so your nose doesn't collide with your partner's nose.

The rest of the muscles in your face and head also play a part in a more involved kiss. For example:

♥ Several muscles move your lips around. Zygomaticus major, zygomaticus minor, and levator labii superioris pull your upper lip and the corners of your mouth upward. Depressor labii inferioris and depressor anguli oris pull the corners of your mouth and your lower lip downward.

♥ If you open your mouth, your lateral pterygoid pulls your

jawbone down. Your massiter, temporalis, and medial pterygoid close your mouth.

♥ Several muscles—your genioglossus, styloglossus, palatoglossus, and hyoglossus—move your tongue if you decide to use it.

Anyone who has ever been kissed knows that the sensations involved aren't confined to the mouth. Your facial nerve carries impulses between your brain and the muscles and skin in your face and tongue. Your brain responds by ordering your body to produce:

♥ Oxytocin, which helps people develop feelings of attachment, devotion, and affection for one another.

♥ Dopamine, which plays a role in the brain's processing of emotions, pleasure, and pain.

♥ Serotonin, which affects a person's mood and feelings.

♥ Adrenaline, which increases heart rate and plays a role in your body's fight-or-flight response.

When you kiss, these hormones and neurotransmitters rush through your body. Along with natural endorphins, they produce the euphoria most people feel during a good kiss. In addition, your heart rate increases and your blood vessels dilate, so your whole body receives more oxygen than it does when you're just standing around. You can also smell the person you're kissing, and researchers have demonstrated a connection between smells and emotions.

KISSING THE BLARNEY STONE

Tourists visiting Ireland often stop by Blarney Castle near Cork to kiss the Blarney Stone. It's said that kissing the stone bestows the kisser with the gift of blarney, or eloquence. But it takes a lot more than just lips. To reach it, people have to lie on their backs, hold a set of handrails, and tip their heads backward until they are essentially upside down.

Your body may also play a role in who you prefer to kiss. Researchers have proven that women prefer men with immune system proteins that are different from their own. In theory, having a baby with someone with different immune proteins can lead to healthier offspring. Scientists believe that a woman may be able to smell these proteins while kissing, and that what she smells may affect whether she finds her partner attractive.

COOTIES FROM KISSING?

Most people know that mouths are germy places. Kissing is directly tied to a few illnesses:

+ Mononucleosis is often called "the kissing disease" because it is carried in saliva and can be spread through kissing.

+ The herpes simplex 1 virus causes cold sores and is easily transmitted through kissing.

+ Although kissing doesn't necessarily cause meningitis, researchers have tracked a correlation between teenagers' number of kissing partners and likelihood of developing the disease.

+ Some researchers theorize that bacteria that cause gastric ulcers may spread through kissing.

+ A person is unlikely to contract the human immunodeficiency virus (HIV) through kissing, but the U.S. Centers for Disease Control and Prevention has reported one case of HIV transmission contracted this way.

SEX ON THE BRAIN: HOW LUST WORKS

I f, as the Bible suggests, love is patient, kind, and unselfish, then lust is the polar opposite. Both forces compel people to couple up in one way or another, but lust doesn't have the time to wait around and woo. It is impatient, brash, and selfish, which are three of the reasons that this innate human behavior has attracted such a ne'er-do-well reputation. In short, it's the vice to love's virtue, associated with short-term sex rather than long-term cuddling.

Going back to antiquity, however, lust arguably has gotten short shrift due to its well-known distinction as one of the seven deadly sins—alongside pride, envy, greed, wrath, gluttony, and sloth. Yet, without lust, the human species would've died off a long, long time ago. Salacious though it may sound, lust is the gatekeeper to love, putting that initial spring in the step and sparkle in the eye when one person is attracted to another.

For that reason, science doesn't characterize lust as an inborn deviance, but as an imperative inertia. Formally defined as the motivational drive to seek out sexual interaction with a member of the same species, lust is the first of three emotional systems—desire, attraction, and attachment—evolved to promote reproduction and long-term mating.

In other words, for all of the waywardness blamed on this controversial four-letter word, lust is actually one of the most practically useful urges of human expression.

Lust in the Brain

Lust originates not in the contours of a shapely calf or a chiseled jawline, but in the hypothalamus, a nugget of neurons in the brain whose function far outweighs its deceptively diminutive size. Situated behind the nose, the hypothalamus directs the pituitary gland to release a range of hormones, including gonadotropin-releasing hormone (GnRH), which has been implicated as a possible human pheromone.

The hypothalamus also oversees the production of a class of hormones called androgens. Testosterone, the starring androgen, along with its chemical cousins dihydrotestosterone and androstenedione, sparks sexual arousal and stirs fundamental physical attraction. For men and women alike, higher levels of testosterone coursing through the body correlate to stronger sex drives and, accordingly, more active sex lives. No wonder that when lust turns into kissing contact, testosterone is exchanged in lovers' saliva.

Lust also puts on an impressive show inside the brain. Repeated cognitive studies have found predictable patterns of arousal in the brain in response to titillating images. An fMRI study conducted in 2002 at the University of Montreal surveyed neurological commotion in males and females while viewing pornographic films. A constellation of brain regions lit up, including sites of visual processing, emotional regulation, and reward. These neurological hotspots include:

- Anterior cingulate (reward).
- Medial prefrontal cortex (sensory processing).
- Orbitofrontal cortex (decision-making).
- Insula (self-awareness).
- Occipitotemporal cortex (visual processing).
- Amygdala (emotional regulation).
- Ventral striatum (reward).

With all of that excitement in the brain, how do humans possibly stand a chance against lusty urges? Fortunately, the brain is also engineered with a safety valve of sorts. In the early 2000s, when University of Montreal neuroscientist Mario Beauregard asked male study participants to mentally resist arousal in the presence of prurient material, their brains engaged parts of the prefrontal cortex involved with self-awareness and behavioral regulation. The right superior frontal gyrus and right anterior cingulate gyrus, in particular, lent a helping hand to delineate between sexual fantasy and reality. That way, the brain serves as a neurological wingman to help the body stand up against lust.

But when all goes according to plan and initial lust leads to a romantic relationship, when does sexual desire develop into outright love?

Lust or Love?

Casually characterized as emotions, love and lust are, more accurately, motivational states. A Venn diagram of the two would certainly show overlap in terms of attraction and the neurological fireworks that burst in the brain when spying the apple of one's eye, but outside of that shared space, what distinguishes the two? The evolved emotional system of lust is a stepping-stone and contributor to love, so how do they function independently?

Advice columns might offer signs to watch out for, such as the amount of time a couple spends in the bedroom together versus the amount of time they spend together elsewhere. And of course, there are those three pivotal words—"I love you"—that can clue a partner in to the other's intentions.

A team of psychologists at the University of Amsterdam

published a trio of studies in 2009 and 2011 that illuminate how love and lust uniquely influence people's thinking patterns. Comparing how sentiments of love and lust foster creativity, lead author Jens Forster and fellow psychologists found that participants primed with feelings of love exhibited broader, long-term thinking processes, lending credence to the romanticized connection between love and artistic expression. Lust, an immediate impulse for sexual satisfaction, inspired more analytic, short-term outlooks.

That kind of love-induced global thinking versus lust-fueled local thinking can be applied to how people perceive their sexual partners. In other words, when thoughts about another person stray from the immediate quandary of Friday night plans to how he or she might fair as a father or mother, a seedling of love may be sprouting.

The love-lust line is also reiterated in the scientifically established phases of long-term mating. By definition, sexual desire—a.k.a. lust—casts a wider net, seeking satisfaction based largely on physical attributes. The segue into genuine affection is marked by specificity, such as craving an emotional union with a special someone rather than just anyone. And for men especially, crossing that border from lust to shared love can come with a statistically higher risk of failure.

EVOLUTION OF LUST

The motivational drive of lust and its accompanying neurobehavioral mechanisms clearly compel people to copulate and reproduce. But some scientists think that those biological underpinnings originally evolved to promote bonding between babies and caregivers. During childbirth, for instance, neurological processes similar to those involved in lust kick in, but with very different results: oxytocin floods a mother's brain, promoting maternal attachment to her baby.

Do Men Lust More?

Is there scientific evidence of a gender gap in lust? Framed solely in terms of sex drive, men unfailingly out-lust women. A Case Western Reserve University meta-analysis of sex drive-related studies from 1996 to 2000 supported that idea by revealing a clear-cut disparity in how often men fantasize about and want to engage in sexual intercourse, compared to women. Moreover, the fMRI analyses mentioned earlier in this section revealed more male brain activity in response to watching erotic films, versus the slightly more sedate female brain-on-porn. Women exhibit robust sex drives, but men may have more piquant arousal patterns and attendant physical urges as a byproduct of sex-drive-revving testosterone.

Being the lustier sex, statistically, comes with its down-side, though. Anyone who's experienced an unrequited crush empathizes with the discomfort that it can produce. The neural circuitry that drives attraction also stimulates lit-eral cravings for a would-be sweetheart's attention and com-pany, triggering a phase called limerance. First described in 1977 by psychologist Dorothy Tennov, limerance encom-passes the aching longing, daydreaming, and fear of rejection that goes along with sustained lust for a potential partner.

While girls might be the ones dotting their i's with hearts and flipping through bridal magazines without a wedding date in sight, boys are far more likely to encounter limerance resulting from unrequited lust. Psychologist Roy Baumeister at Florida State University estimates that men are more likely than women—by a three-to-two ratio—to pine away for people who won't reciprocate.

On the bright side, those lusty letdowns could be nature's way of preserving love as the intimate, profound motivational state that humans treasure so dearly. After all, if every lustful whim could be satisfied, people would probably never sit down and stay a while.

THE DOPE ON LOVE DRUGS: HOW APHRODISIACS WORK

M ost of us have heard the talk about oysters and chocolate, and maybe you've read an article about the stimulating effects of ginseng. But garlic, licorice, and cucumber?

The "love" industry is booming, in case you hadn't noticed from your e-mail inbox lately. Spammers have hit on aphrodisiacs as a prime seller, and nutritional-supplement manufacturers are also getting in on the action. What's usually missing is the clinical research to prove these "aphrodisiacs" work. Can certain foods, drugs, and scents really increase sexual desire? Millions of people swear they can, but the scientific evidence is still in question.

In this section, we'll find out what types of things people believe are aphrodisiacs and whether or not there really is anything to those beliefs.

The Basic Idea

By definition, aphrodisiacs are elements that evoke or stimulate sexual desire. Companies that produce drugs or concoctions that claim to enhance your sex life often mislabel these supplements as aphrodisiacs. In order to be a true aphrodisiac, they would have to *create desire*—not improve performance and ability. Viagra, for example, is not an aphrodisiac.

Before we can determine if something works, we have to understand what it would have to do to work. In the case of aphrodisiacs, what happens in the body and brain when we are sexually excited? For both men and women, it all boils down to hormones—specifically testosterone. In other words, our sex drive is controlled by our hormone levels, with testosterone being the key. If the balance is off, things may not function as they should. When it's right, everything falls into place.

A chain reaction begins when we see, hear, feel, think, smell, or otherwise encounter something sexually stimulating. The process looks something like this:

- First, signals are sent from the limbic lobe of the brain via the nervous system to the pelvic region. These signals tell the blood vessels to dilate.
- This dilation creates an erection (in both men and women—the female erectile tissues are found in the clitoris and the region around the vaginal entrance).
- The vessels then close so those erectile tissues stay erect. This erection is accompanied by rapid heart rate.

♥ At the same time, our brains are releasing norepineph-rine and dopamine, neurotransmitters that tell our bodies that this is good and pleasurable. (For more about the chemical processes involved with love, read the section titled "Head Over Heels: How Love Works.")

If we don't have enough testosterone, our interest in sex can dwindle. Other factors including stress, fatigue, and depression can also have a big effect on sexual interest.

UNEXPECTED SIDE EFFECTS FROM DRUGS

Aphrodisiacs with the most probable effects are those that have been discovered because people reported increased sexual activity as a side effect of a pre-scribed drug. Many drug companies are using this unexpected side effect to develop drugs that address the lack of sexual desire some people experience.

You Are What You Eat: What Aphrodisiacs Do

Experts say that aphrodisiacs can work in two ways. Some create sexual desire by working on the mind, while others create desire by affecting parts of the body. For example, something that increases blood flow in the sex organs might simulate the feelings of sexual intercourse and have the effect of creating desire. Likewise, there are things that can make our bodies produce more of the chemicals associated with sexual desire.

Something that lowers inhibitions in the mind, such as alcohol or marijuana, might also create (or allow) the desire to have sex. Sometimes just thinking that something is an aphrodisiac makes it appear to work as one. There are also things that quell desire. These are called anaphrodisiacs.

Researchers are finding that some foods, herbs, and other supplements do stimulate production of hormones or other chemicals that affect our libidos. What they don't know is whether those chemicals are produced in a high enough quantity for us to really notice the difference. There isn't much hard research in the area, primarily because libido is a somewhat difficult thing to study, especially in women, whose arousal doesn't produce as much of a physical reaction as in men and therefore can't be as easily measured.

IF IT LOOKS LIKE A SEX ORGAN, IT MUST DO SOMETHING...

What would make someone think that something like a rhino horn could have any power as an aphrodisiac?

Well, for one thing, a rhino horn bears a resemblance to an erect penis. A resemblance to sexual organs like this has often made people think that something must have sexual powers. Vegetables such as carrots, asparagus, and cucumbers have all been associated with aphrodisiacs, even if their chemical makeup shows no relationship (although some have been shown to have chemical characteristics that could possibly contribute to improved sexual desire).

Aside from seeing a resemblance to sex organs, people throughout history have made aphrodisiac associations with animals that are known to be virile and prolific reproducers. Rabbits, tigers, goats, and bulls, for instance, have reputations for prolific reproduction, strength, and virility. Historically, people ingested the sex organs of these animals to achieve an aphrodisiac effect or to enhance sexual performance, or both. Cave drawings depicted hunters eating the testicles of animals they killed, and the belief is that they hoped to take on the characteristics of that animal.

According to the U.S. Food and Drug Administration, aphrodisiacs have no scientific basis and are simply myth.

While this may be true, many people swear by the effects of certain foods, herbs, or minerals. The following is a list of foods that reportedly act as aphrodisiacs. Some are said to be aphrodisiacs simply because of their shape and some because of their aromas, while others claim a chemical basis for their "love" powers. *Note: This is by no means a complete list and, unless otherwise noted, there is no readily available research to back up the claims.*

ANISEED

The ancient Greeks and Romans believed that you could increase desire by sucking on anise seeds. Aniseed, also known as anise, does include estrogenic compounds (female hormones), which have been reported to induce similar effects to testosterone.

AVOCADO

The avocado tree was called a "testicle tree" by the Aztecs because its fruit hangs in pairs on the tree, resembling the male testicles. Its supposed aphrodisiac value is based on this resemblance.

BANANAS

In addition to the phallic shape of the banana itself, the banana flower has a phallic shape. Bananas are rich in potassium and B vitamins, which are said to be necessary for sex-hormone production.

BASIL (SWEET BASIL)

For centuries, people said that basil stimulated the sex drive and boosted fertility, as well as producing a general sense of well-being. The scent of basil was said to drive men wild—so much so that women would dust their breasts with dried and powdered basil. Basil is one of the many reported aphrodisiacs that may have the property of promoting circulation.

CARDAMOM

Cardamom is an aromatic spice. Certain cultures deem it a powerful aphrodisiac and also claim it is beneficial in treating impotence. It is high in cineole, which can increase blood flow in areas where it is applied.

CHOCOLATE

Chocolate has forever been associated with love and romance. It was originally found in the South American rain forests. The Mayan civilizations worshipped the cacao tree and called it "food of the gods." Rumor has it that the Aztec ruler Montezuma drank fifty goblets of chocolate each day to enhance his sexual abilities.

Researchers have studied chocolate and found it to contain phenylethylamine and serotonin, which are both "feel good" chemicals. They occur naturally in our bodies and are released by our brains when we are happy or feeling loving or passionate. They produce a euphoric feeling, like when you're in love.

In addition to those two chemicals, researchers at the Neurosciences Institute in La Jolla, California, say that chocolate may also contain substances that have the same effect on the brain as marijuana, including a neurotransmitter called anandamide. The amount of anandamide in chocolate is not enough to get a person "high" like marijuana, but it could be enough to contribute to the good feelings that serotonin and phenylethylamine produce. Does that mean it increases sexual desire? Probably not—but if it makes you feel good, it might lower your inhibitions so that you're more receptive to suggestion.

CHILI PEPPERS

Eating chili peppers generates physiological responses in our bodies (for example, sweating, and increased heart rate and circulation) that are similar to those experienced when having sex. The capsaicin that peppers contain is responsible for the effects and is also a good pain reliever. Another reported effect of eating large quantities of chili peppers is an irritation of the genitals and urinary tract that could feel similar to sexual excitement.

CUCUMBERS

Aside from its phallic shape, the cucumber has a scent that is believed to stimulate women by increasing blood flow to the vagina.

FIGS

The fig is another fruit that claims aphrodisiac qualities based on its appearance. A cut-open fig is thought to look similar to female sex organs.

GARLIC

Long ago, Tibetan monks were not allowed to enter the monastery if they had been eating garlic because of its reputation for stirring up passions. Garlic increases circulation.

GINGER

People have deemed ginger root an aphrodisiac for centuries because of its scent and because it stimulates the circulatory system.

HONEY

In medieval times, people drank mead, a fermented drink

made from honey, to promote sexual desire. In ancient Persia, couples drank mead every day for a month (known as the "honey month"—a.k.a. "honeymoon") after they married to get in the right frame of mind for a successful marriage. Honey is rich in B vitamins (needed for testosterone production) as well as boron (helps the body metabolize and use estrogen). Some studies have suggested that honey may also enhance blood levels of testosterone.

LICORICE

In ancient China, people used licorice to enhance love and lust. The smell appears to be particularly stimulating. Alan R. Hirsch, MD, neurological director of the Smell and Taste Treatment and Research Foundation in Chicago, conducted a study that looked at how different smells stimulated sexual arousal. He found that the smell of black licorice increased the blood flow to the penis by 13 percent. When combined with the smell of doughnuts, that percentage jumped to 32.

NUTMEG

In ancient China, women prized nutmeg an aphrodisiac. Researchers have found it to increase mating behaviors in mice, but no evidence proves that the same happens in humans. In large quantities, nutmeg can produce a hallucinogenic effect.

OYSTERS

Romans documented oysters as aphrodisiacs in the second century AD. The bivalves are known to be high in zinc, which has been associated with improving sexual potency in men. (An additional hypothesis is that the oyster resembles the female genitals.) Mussels, clams, and oysters have been found to contain D-aspartic acid and NMDA (N-methyl-D-aspartate), compounds that may be effective in releasing sex hormones like testosterone and estrogen. Scientists have not determined whether there are enough of those compounds in the shellfish to make any difference.

PINE NUTS

People have been using pine nuts to stimulate the libido since medieval times. Like oysters, pine nuts are high in zinc. They have been used for centuries to make up love potions. The Arabian medical scholar Galen recommended eating one hundred pine nuts before going to bed.

♥ ♥ ♥ ♥ ♥

Now let's take a look at some aphrodisiac non-foods and how they're supposed to accomplish their effects.

Aphrodisiac Supplements and What They Claim to Do

Like the aphrodisiac foods listed before, some of the aphrodisiac ingredients found in supplements may have research to back their claims, but most do not.

ARGININE

Arginine is an amino acid found in meat, nuts, eggs, coconut milk, and cheese. It forms nitric oxide in the body, which increases blood flow to the genitals. When combined with other supplements, arginine is said to enhance sexual desire in women.

EPIMEDIUM

The epimedium herb has been proven to improve the sexual function of male animals in experiments, according to pharmacognosist Albert Leung, PhD, and Arkansas herbalist Steven Foster. It acts somewhat as an androgen (sex hormone) and might stimulate sexual desire in women who are androgen-deficient.

FENNEL

Fennel is reported to increase the libido of both male and female rats. Fennel has compounds that mimic the female hormone estrogen. However, in doses greater than about a teaspoon, it can be toxic!

GINSENG

Ginseng is another long-touted aphrodisiac. Researchers reported in the *Journal of Urology* that "mean International Index of Erectile Function scores were significantly higher in patients treated with Korean red ginseng than in those who received placebo." Animal studies have not shown that ingesting ginseng has an immediate effect on testosterone levels, but the ginseng may trigger other mechanisms that lead to increased performance and libido.

RHINO HORN

Rhino horn is primarily fibrous tissue with fairly large amounts of calcium and phosphorus. Since low levels of these minerals can lead to weakness and general fatigue, taking large doses of these elements could increase stamina if levels were low to begin with. It's understandable, then, how rhino horn could have seemed historically to be an aphrodisiac (in addition to its resemblance to an erect penis). People who didn't have deficiencies of those minerals wouldn't have seen the same effect.

SPANISH FLY

Probably one of the most famous aphrodisiacs is Spanish fly. It is made from a beetle that secretes an acid-like juice, called cantharidin, from its leg joints when threatened. Because it would be more difficult to remove just the juice, the entire beetle is dried and crushed to produce the powder. When Spanish fly powder is ingested, the body excretes the cantharidin in the urine. This causes intense irritation and burning in the urogenital tract, which then leads to itching and swelling of the genitals.

This swelling and burning was once assumed to be sexual arousal and led to the belief that Spanish fly had aphrodisiac qualities. But cantharidin is highly toxic. The kidneys suffer inflammation as well and can be permanently damaged. Spanish fly can cause severe gastrointestinal disturbances, convulsions, and even death.

YOHIMBE

Yohimbe is used both as an herbal aphrodisiac and in a prescription drug used for erectile dysfunction in men. It comes from bark stripped from a West African evergreen tree. Yohimbine, the primary active ingredient of yohimbe, blocks alpha-2 adrenergic receptors and increases dilation of blood vessels, which are both involved in achieving and maintaining an erection. The herbal form of yohimbe can be dangerous if taken in the wrong quantities.

WHY ARE RHINOS ENDANGERED?

Contrary to popular belief, rhinos are not endangered because of the use of their horns as aphrodisiacs. According to the World Wildlife Fund, rhino horn is a valuable ingredient in traditional Chinese medicine and is used effectively to reduce fevers and to treat delirium, high blood pressure, and other ailments. The use of rhino horn is now banned throughout most of Asia, but hundreds of rhinos were killed each year prior to the ban, and many are still killed illegally. In addition to their use in medicines, rhino horns are carved to make the handles of ceremonial daggers.

Other Types of Aphrodisiacs

Scents have a powerful effect when it comes to romance. The memory of the scent of a romantic partner can stay with us long after the romance is gone—so much so that when we run across the scent years later, we're immediately taken back with a flood of feelings and memories. Does that mean that scent is an aphrodisiac?

In a way, yes, in that scent can evoke desires—but typically not in an otherwise unwilling partner. For example, Alan R. Hirsch, MD, neurological director of the Smell and

Taste Treatment and Research Foundation in Chicago, conducted a study that looked at how different smells stimulated sexual arousal. He found that several scents were effective—some more than others.

The smell of cheese pizza, for instance, increased blood flow to the penis by 5 percent, buttered popcorn by 9 percent, and lavender and pumpkin pie each by 40 percent. For women, lavender and pumpkin pie also had a stimulating effect. However, the smell of Good & Plenty candy (licorice) combined with the scent of cucumber created the greatest increase in blood flow to the vagina.

Human pheromones, which still carry some weight in the field of love research, may actually create sexual interest. The word "pheromone" comes from the Greek words *pherein* and *hormone*, meaning "excitement carrier."

In the animal world, pheromones are individual scent "prints" found in urine or sweat that dictate sexual behavior and attract the opposite sex. They help animals identify each other and choose a mate with an immune system different enough from their own to ensure healthy offspring. They have a special organ in their noses called the vomeronasal organ (VNO) that detects this odorless chemical.

In 1986, scientists at the Monell Chemical Senses Center in Philadelphia and its counterpart in France discovered the existence of human pheromones in human sweat.

A human VNO has also been found in some, but not all, people. Even if the VNO isn't present in all of us—and may not be working in those who do have it—there is still evidence that smell is an important aspect of love. (Note the booming perfume industry.)

MUSIC

Music can set the mood, carry the mood, or ruin the mood. What appears to be the most effective element of music is the memory we associate with it. If you have fond memories of slow dancing to a special song with someone you loved in the past, it's a good bet that same song will have an effect on you later in life. After all, the dance is a bit of a mating ritual that most of us have experienced at some point in our lives.

EXERCISE

Not only is it good for your health, but it's also good for your sex life. According to Discovery Health, the aphrodisiac qualities of exercise are associated with the endorphins released in the brain with vigorous activity—like the runner's high. Endorphins are those "feel good" chemicals. Other aphrodisiac effects of exercise come from exercises that increase blood flow to the genitals. These exercises position the body in various ways that stimulate blood flow and can improve sexual abilities and desire.

Exercise and building muscle mass usually increases testosterone levels, too, which may be another reason why exercise increases sex drive.

Is It All in Our Heads?

So is there really something to aphrodisiacs, or is it all in our heads? Some say the latter: in other words, it's a placebo effect. If we think something is going to put us in the mood for love, we'll find ourselves there. The aphrodisiac we consumed may or may not have anything to do with it.

Are we already halfway there simply by thinking something is going to work? The answer, according to most sources, is yes. Some studies have shown that agents that appear to work amazingly well one time might have no effect the next time, even on the same people. This leads most scientists to believe that aphrodisiacs have a greater effect in our heads than in other parts of our anatomies. After all, the body's most powerful sex organ is the brain. If your head isn't in the right place, nothing is going to happen.

Historic claims about the aphrodisiac effects of certain foods or supplements may have held more truth at the time than they do today because overall nutrition wasn't as good then. Taking or eating something that was rich in nutrients would have had a more profound effect on overall health, which in turn would affect sexual desire, making it appear that the food, herb, or supplement had aphrodisiac qualities. People are simply healthier now than in the past, so it's more difficult to see the effects of particular nutrient-rich foods.

THE BIG O: WHAT HAPPENS IN THE BRAIN DURING AN ORGASM?

For many people, sex is just a precursor to this intensely pleasurable event. What exactly is going on inside when you lose control?

Because we're all so different, coming up with a universal description of an orgasm is impossible. The one thing that most people can agree on is that it's an incredibly, intensely pleasurable experience.

So what is it? When in doubt, go to the dictionary. The *Oxford English Dictionary* defines an orgasm as "a sudden movement, spasm, contraction, or convulsion […] a surge of sexual excitement." *Merriam-Webster* gets more descriptive, stating that it's "an explosive discharge of neuromuscular tensions at the height of sexual arousal that is usually accompanied by the ejaculation of semen in the male and by vaginal contractions in the female." The famous sex researcher Alfred Kinsey once said that an orgasm "can be likened to the crescendo, climax, and sudden stillness achieved by an orchestra of human emotions…an explosion of tensions, and to sneezing."

Kinsey's comparison to sneezing might be debatable, but other than that, all of these definitions are basically correct, and are just a few of the many different attempts to describe exactly what it means to have an orgasm. Nearly every aspect of the orgasm—what's required to have one, why some people can't seem to achieve one, why we have them at all—has been the subject of much research and debate. What happens to the body during an orgasm is fairly well-known, and it's no surprise that the brain plays a big part in reaching one. But researchers are still in the process of figuring out exactly what's happening in the brain during an orgasm. Let's start by looking at the messages that the body sends to the brain.

Orgasms and Nerves

Without nerves sending impulses back to the spinal cord and brain, an orgasm wouldn't be possible. Just like any other area of the body, the genitalia contain different nerves that send information to the brain to tell it about the sensation that's being experienced. This helps to explain why the sensations are perceived differently depending on where someone is being touched. A clitoral orgasm, for example, differs from a vaginal orgasm because different sets of nerves are involved.

All of the genitalia contain a huge number of nerve endings (the clitoris alone has more than eight thousand of them), which are, in turn, connected to large nerves that run up through the body to the spinal cord. (The exception is the vagus nerve, which bypasses the spinal cord.) They perform many other functions in the body in addition to providing the nerve supply—and therefore feedback to the brain—during sexual stimulation. Here are the nerves and their corresponding genital areas:

- **Hypogastric nerve**—Transmits from the uterus and the cervix in women and from the prostate in men.
- **Pelvic nerve**—Transmits from the vagina and cervix in women and from the rectum in both sexes.
- **Pudendal nerve**—Transmits from the clitoris in women and from the scrotum and penis in men.
- **Vagus nerve**—Transmits from the cervix, uterus, and vagina.

The role of the vagus nerve in orgasms is a new discovery, and much is still unknown about it. Until the twenty-first century, researchers didn't know that the long, wandering nerve passed through the pelvic region at all.

Since most of those nerves are associated with the spinal cord, it would stand to reason that a person with a severed spinal cord wouldn't be able to have an orgasm. And for a very long time, that's what people with these types of injuries were told. However, at least one study has shown that people with spinal cord injuries—even paraplegics—can reach orgasm.

Psychology professors Barry Komisaruk and Beverly Whipple of Rutgers University conducted a study on women with severed spinal cords in 2004. They discovered that these women could feel stimulation of their cervixes and even reach orgasm, although there was no way their brain could be receiving information from the hypogastric or pelvic nerves. How was this possible? An MRI scan of the women's brains showed that the region corresponding to signals from the vagus nerve was active. Because the vagus bypasses the spinal cord, the women were still able to feel cervical stimulation.

So during sexual stimulation and orgasm, different areas of the brain receive all of this information about exactly what's happening—and that what's happening is very enjoyable. Before all this research, we had no way of knowing what was happening in the brain at the exact moment of orgasm.

Pleasure Center of the Brain: Light It Up

You may have heard that the brain has a pleasure center that lets us know when something is enjoyable and reinforces our desire to perform that same pleasurable action again. This is also called the reward circuit and responds to all kinds of pleasure—from sex to laughter to certain types of drug use. Some of the brain areas impacted by pleasure include:

- ♥ **Amygdala**—Regulates emotions.
- ♥ **Nucleus accumbens**—Controls the release of dopamine.
- ♥ **Ventral tegmental area (VTA)**—Actually releases the dopamine.
- ♥ **Cerebellum**—Controls muscle function.
- ♥ **Pituitary gland**—Releases beta-endorphins, which decrease pain; oxytocin, which increases feelings of trust; and vasopressin, which increases bonding.

Although scientists have studied the pleasure center for a long time, there hadn't been much research about how it relates to sexual pleasure, especially in women. During the late 1990s and the following decade, a team of scientists at the University of Groningen in the Netherlands conducted several studies of both men and women to determine brain activity during sexual stimulation. The team used PET scans to illustrate the different areas of the brain that would light up and shut off during sexual activity. In all of the tests, the subjects were scanned while resting, while being sexually stimulated, and while having an orgasm.

Interestingly, the researchers didn't discover many differences between men's and women's brains when it comes to sex. In both, the brain region behind the left eye, called the lateral orbitofrontal cortex, shuts down during orgasm. Janniko R. Georgiadis, one of the researchers, said, "It's the seat of reason and behavioral control. But when you have an orgasm, you lose control." Neuroscientist and fellow researcher Gert Holstege stated that the brain during an orgasm looks much like the brain of a person taking heroin, saying that "95 percent is the same."

There are some differences, however. When a woman has sex, a part of the brain stem called the periaqueductal gray (PAG) is activated. The PAG controls the fight-or-flight response. Women's brains also showed decreased activity in the amygdala and hippocampus, which deal with fear and anxiety. The team theorized that these differences existed because women have more of a need to feel safe and relaxed in order to enjoy sex. In addition, the area of the cortex associated with pain was activated in women, which shows a distinct connection between pain and pleasure.

The studies also showed that although women may be able to fool their partners into thinking they've had an orgasm, their brains show the truth. When women were asked to fake an orgasm, their brain activity increased in the cerebellum and other areas related to controlling movement. The scans didn't show the same brain activity of a woman during an actual orgasm.

But what about people who can't reach orgasm at all?

Neither Here nor There: Anorgasmia and Non-Genital Orgasms

In some cases, we know what causes anorgasmia (the inability to reach orgasm). Drugs like Celexa, Zoloft, and Paxil—known as SSRIs, or selective seratonin reuptake inhibitors—are often used to treat depression, anxiety, and other mental illnesses. Like most drugs, however, they can have side effects. For some people, this includes sexual ones, including anorgasmia. But why?

SSRIs can decrease the brain's production of dopamine, the neurotransmitter that provides pleasurable feelings and reinforces a person's desire to once again perform the action that brought him or her pleasure. Sometimes the problem goes away on its own, or it can be resolved by switching to a different antidepressant or taking another drug in addition to the SSRI.

However, a small number of people experience post-SSRI sexual dysfunction (PSSD) that lasts for days, weeks, months, or even years after discontinuing use of an SSRI. The cause of this dysfunction isn't understood, as stopping the SSRI allows dopamine production to return to normal. But various studies are continuing to research this issue in order to help women who are anorgasmic.

Perhaps more unusual sounding than anorgasmia is the concept of orgasms that have nothing to do with the genitalia.

Some people can orgasm from being touched in other places on the body, such as the nipples. In this case, researchers believe that the sensations in the nipples are transmitted to the same areas of the brain that receive information from the genitals. However, people have also reported actually feeling orgasms in other parts of their bodies, including their hands and feet.

Several people even described having orgasms in limbs that were no longer there. One reason for that may be the layout of the cortical homunculus, a map that shows how different places of the brain's sensory and motor cortices correspond to the organs and limbs of the body. A person who feels an orgasm in a phantom foot, for example, may have experienced a remapping of the senses because the foot is located next to the genitals in the homunculus. The foot is no longer there to provide sensation, so the area for genital sensation took over the space.

We now know more about how orgasms impact the brain than ever before, but there's still a lot that we don't know. For example, scientists are still debating the evolutionary reason behind the female orgasm. But it's probably safe to say that most people aren't too concerned about the "why"—they'd prefer to focus on the who, what, and when of sex.

2

MEETING
YOUR MATCH

IS LOVE AT FIRST SIGHT POSSIBLE?

How many movies have you seen where a glimpse across the room is enough to convince a protagonist that the search for love is over? Silly, right? Some scientists don't think so. Is love at first sight possible?

Falling in Love in Three Minutes or Less

Helen Fisher, a prominent anthropologist known for her research on attraction and love, believes three minutes is all you need to know whether someone will be in your life for a while. To understand her theory, we have to travel back in time to the days of early humans. Our ancestors lived shorter lives than we do, and it was important in their brief time on Earth to mate and produce a healthy child so that the race would live on. For this reason, they had to size up potential mates quickly, just as they had to quickly size up whether a stranger was friend or foe. Fisher believes our evolutionary past wired our brains so that we know rather quickly whether we might want to mate with someone (even if we're not even looking to have a child).

So what are we considering in those three minutes? Many scholars speak of the concept of a "lovemap," a laundry list of traits that we want in a partner (more on this in the next section). This means that when you told a girlfriend that your next boyfriend needed to be tall and have a sense of humor, you were actually working on a lovemap. But while you may have some ideas about what you find attractive in a potential paramour, these ideals of beauty were likely influenced by those evolutionary ancestors again.

Men and women both wanted to ensure that their children would live and pass on their genes, so they needed to be sure that the other party was bringing the best genetic makeup to the table. We often signal our physical and reproductive health with traits like a certain waist-to-hip ratio or a symmetrical face; scientists have found that these qualities are universally attractive to others. And when you check out a guy's chin or a lady's lovely eyes, you're actually looking at traits that are shaped by the amounts of testosterone and estrogen in their bodies, respectively, which also indicate reproductive fitness. So when we comment on someone's hotness, we're actually commenting on ancient ideals of fertility.

So we can tell fairly quickly whether someone will give us a cute, healthy baby. But is that love, or just lust? Fisher points out that the sections of the brain that respond to love and lust are different, though they can light up at the same time. In a study conducted at Syracuse University, researchers found that the hormones associated with love, rather than lust, can flood the brain in one-fifth of a second. It seems to indicate that our brain can start feeling amorous rather quickly, but in the next section, we'll consider more elements of the lovemap and what else might be going on in that 0.2-second to three-minute time span.

Your Brain's Love Checklist

You might be saying to yourself that love is more than a physical attraction to someone, and it's true that other things are going on when you first encounter a potential mate. For one thing, without you sensing it, your brain is sizing up this person against all other past loves. If you're taking notice of a guy with a baseball cap, your brain may nudge you with, "Hey, remember that last man you dated who wore baseball caps? That didn't go so well."

Or, your brain may be measuring visual cues against stereotypes about socioeconomic status. If you're spying someone with a briefcase and a business suit, your brain is helping you conclude that this person may work a lot, but at least they can afford a few nice dinners every now and then. Though this insight may disappoint romantics, your brain is only trying to protect you and your assets so that you don't date a broke man who wears baseball caps like your last dud of a boyfriend.

While you're making small talk, you're sizing up his or her voice. A male who speaks deeply and quickly, for example, is likely to be rated as better-looking and more highly educated by women around him, according to some studies. And of course, if your date is saying things that jibe with your worldview, then you're going to be further besotted.

Though we often hear that opposites attract, scientists say we're far more likely to practice assortative mating, which is partnering off with people who are similar to us. Successful couples may share the same religious values and tax bracket, and they tend to be "in the same league," looks-wise. Yes, what you learned in high school is true—the pretty people tend to stick with their kind. One study found that people tend to choose people who have the same level of body fat.

Of course, we don't want someone who's too much like us genetically. Remember our ancestors' mandate to find someone who could make a baby with the best chance of survival? As previously mentioned, that's why researchers also believe that smell is involved when we're sizing up the opposite sex.

LOVE AT FIRST SIGHT—FOR THE BLIND

In a 2009 BBC article, blind writer Damon Rose wrote about how the sightless fall in love. As you might imagine, a good voice has a lot to do with it. But Rose pointed out that blindness doesn't exempt people from shallowness. He provided several examples of blind people who were enamored with a certain person—until someone else alerted the blind person that the companion was unattractive. Not that explaining your attractiveness will help you win a blind person's heart. Rose said he found people who talked about their looks egotistical, which makes us wonder: are blind people the pickiest daters of all?

And to make everything more complicated, it might all come down to what time of the month you spy a lovely lady or a handsome gent. There's evidence that women become more attuned to certain traits in men during the most fertile times in their menstrual cycles. Specifically, women tend to respond more strongly to potential suitors when they're ovulating, and men, in turn, tend to find women more attractive during the same period, even when the men don't know the lady's cycle. One interesting study even found that exotic dancers tended to receive much higher tips at their most fertile points of the month.

Perhaps you're seeing how difficult it can be to fall in love at first sight—the person in question has to possess the right genes and look like someone we could be with, according to our mental lovemaps. But just as important as finding a person who looks *like* us, though, is finding a person who looks *at* us.

In a study published in the *Proceedings of the Royal Society*, one researcher likened love at first sight to narcissism, because we're most attracted to someone who happens to be looking at us. Again, this has evolutionary roots, as we shouldn't spend time chasing a mate who's not interested, but it's narcissistic because the person we tend to look at, of course, looks like us. It's like falling in love with your own image in the mirror.

HEAD OVER HEELS: HOW LOVE WORKS

I f you've ever been in love, you've probably at least considered classifying the feeling as an addiction. And guess what? You were right. Scientists are discovering that the same chemical process that takes place with addiction takes place when we fall in love.

Love is a chemical state of mind that's part of our genes and influenced by our upbringing. We are wired for romance in part because we are supposed to be loving parents who care diligently for our helpless babies.

What Makes Us Fall in Love?

We all have a template for the ideal partner buried somewhere in our subconscious. This lovemap decides which person in that crowded room catches our eye. But how is this template formed?

APPEARANCE

Many researchers have speculated that we tend to go for members of the opposite sex who remind us of our parents.

Some have even found that we tend to be attracted to those who remind us of ourselves. In fact, cognitive psychologist David Perrett, at the University of St. Andrews in Scotland, did an experiment in which he morphed a digitized photo of the subject's own face into a face of the opposite sex. Then, he had the subject select from a series of photos which one he or she found most attractive. According to Perrett, his subjects always preferred the morphed version of their own face (and they didn't recognize it as their own).

PERSONALITY

Like appearance, we tend to form preferences for those who remind us of our parents (or others close to us through childhood) because of factors like their personality, sense of humor, and likes and dislikes.

PHEROMONES

As previously discussed, pheromones, the chemicals that give off a person's unique smell, can play a big part in a lovemap. An experiment was conducted where a group of females smelled the unwashed T-shirts of a group of sweaty males, and each had to select the one to whom she was most "attracted." Just like in the animal world, the majority of the females chose a shirt from the male whose immune system was the most different from their own.

STARING INTO EACH OTHER'S EYES

Professor Arthur Aron, of Stony Brook University, has studied what happens when people fall in love and has found that simply staring into each other's eyes has tremendous impact. In an experiment he conducted, Aron put strangers of the opposite sex together for ninety minutes and had them discuss intimate details about themselves. He then had them stare into each other's eyes for four minutes without talking. The results? Many of the subjects felt a deep attraction for their partner after the experiment, and two even ended up getting married six months later.

⟩ Stages of Love: Lust and Attraction

There are three distinct types or stages of "love":

1. Lust, or erotic passion.
2. Attraction, or romantic passion.
3. Attachment, or commitment.

When all three of these happen with the same person, you have a very strong bond. Sometimes, however, the one we lust after isn't the one we're actually in love with.

LUST

When we're teenagers, just after puberty, estrogen and testosterone become active in our bodies for the first time and create the desire to experience "love." These desires, a.k.a. lust, play a big role both during puberty and throughout our lives. Lust and romantic love are two different things caused by different underlying substrates, according to an article by psychology professor Lisa Diamond in *Current Directions in Psychological Science*.

Lust evolved for the purpose of sexual mating, while romantic love evolved because of the need for infant and child bonding. So even though we often experience lust for our romantic partner, sometimes we don't—and that's okay. Or, maybe we do, but we also lust after someone else. According to Diamond, that's normal.

Sexologist John Money draws the line between love and lust in this way: "Love exists above the belt, lust below. Love is lyrical. Lust is lewd." Pheromones, looks, and our own learned predispositions for what we look for in a mate also play an important role in who we lust after. Without lust, we might never find that special someone. But, while lust keeps us "looking around," our desire for romance leads us to attraction.

ATTRACTION

While the initial feelings may (or may not) come from lust, what happens next—if the relationship is to progress—is attraction. When attraction, or romantic passion, comes into play, we often lose our ability to think rationally—at least when it comes to the object of our attraction. The old saying "love is blind" is really accurate at this stage.

We are often oblivious to any flaws our partner might have. We idealize them and can't get them off our minds. This overwhelming preoccupation and drive is part of our biology. (We'll go deeper into the chemicals involved in attraction in "The Chemistry of Love," on the next page.) In this stage, couples spend many hours getting to know each other. If this attraction remains strong and is felt by both of them, then they usually enter the third stage: attachment.

ATTACHMENT

The attachment, or commitment, stage is love for the duration. You've passed fantasy love and are entering into real love. This stage of love has to be strong enough to withstand many problems and distractions. Studies by University of Minnesota social psychologist Ellen Berscheid and others have shown that the more we idealize the one we love, the stronger the relationship during the attachment stage.

Psychologists at the University of Texas in Austin have

come to the same conclusion. They found that idealization appears to keep people together and keep them happier in marriage. "Usually, this is a matter of one person putting a good spin on the partner, seeing the partner as more responsive than he or she really is," says Ted Huston, the study's lead investigator. "People who do that tend to stay in relationships longer than those who can't or don't." Playing a key role in this stage are oxytocin, vasopressin, and endorphins, which are released when having sex (more on this later).

Let's find out more about the chemistry of love.

"Be Still, My Beating Heart": The Chemistry of Love

There are a lot of chemicals racing around your brain and body when you're in love. Researchers are gradually learning more about the roles these chemicals play both when we are falling in love and when we're in long-term relationships. Of course, estrogen and testosterone play a role in the sex drive area. Without them, we might never venture into the "real love" arena.

That initial giddiness that comes when we're first falling in love includes a racing heart, flushed skin, and sweaty palms. Researchers say this is due to the dopamine, norepinephrine, and phenylethylamine we're releasing. Dopamine is thought to be the "pleasure chemical," producing a feeling of bliss. Norepinephrine is similar to adrenaline and produces the racing heart and excitement. Together, these two chemicals produce elation, intense energy, sleeplessness, craving, loss of appetite, and focused attention, according to Helen Fisher, anthropologist and well-known love researcher from Rutgers University. She also says, "The human body releases the cocktail of love rapture only when certain conditions are met and…men more readily produce it than women, because of their more visual nature."

Researchers are using fMRI to watch people's brains when they look at a photograph of their object of affection. Fisher says that what they see in those scans during that "crazed, can't-think-of-anything-but stage of romance"— the attraction stage—is the biological drive to focus on one person.

The scans showed increased blood flow in areas of the brain with high concentrations of receptors for dopamine, which is associated with states of euphoria, craving, and addiction. High levels of dopamine are also associated with norepinephrine, which heightens attention, short-term memory, hyperactivity, sleeplessness, and goal-oriented behavior. In other words, couples in this stage of love focus intently on the relationship and often on little else.

Another possible explanation for the intense focus and idealizing that occurs in the attraction stage comes from researchers at University College London: They discovered that people in love have lower levels of serotonin and also that neural circuits associated with the way we assess others are suppressed. These lower serotonin levels are the same as those found in people with obsessive-compulsive disorders, possibly explaining why those in love "obsess" about their partner.

Chemical Bonding

In romantic love, when two people have sex, oxytocin is released, which helps bond the relationship. The hormone oxytocin has been shown to be "associated with the ability to maintain healthy interpersonal relationships and healthy psychological boundaries with other people," according to researchers at the University of California, San Francisco. When oxytocin is released during orgasm, it begins creating an emotional bond—the more sex, the greater the bond. Oxytocin is also associated with mother-infant bonding, uterine contractions during labor in childbirth, and the "let down" reflex necessary for breastfeeding.

Vasopressin, an antidiuretic hormone, is another chemical that has been associated with the formation of long-term, monogamous relationships. (See "Are We Alone in Love?") Fisher believes that oxytocin and vasopressin interfere with the dopamine and norepinephrine pathways, which might explain why passionate love fades as attachment grows.

Endorphins, the body's natural painkillers, also play a key role in long-term relationships. They produce a general sense of well-being, including feeling soothed, peaceful, and secure. Like dopamine and norepinephrine, endorphins are released during sex, as well as during physical contact, exercise, and other activities. According to Michel Odent of London's Primal Health Research Center, endorphins induce a "drug-like dependency."

LOVE JUNKIES

Some people may be addicted to the love "high." They need that amphetamine-like rush of dopamine, norepinephrine, and phenylethylamine. Because the body builds up a tolerance to these chemicals, it begins to take more and more to give love junkies that high. They go through relationship after relationship to get their fix.

⋛ The Long Haul?

What about when that euphoric feeling is gone? The speed at which courtship progresses often determines the ultimate success of the relationship, according to psychologist Ted Huston at the University of Texas. What he and fellow researchers found was that the longer the courtship, the stronger the long-term relationship.

However, feelings of passionate love do lose their strength over time. Studies have shown that passionate love fades quickly and is nearly gone after two or three years. The chemicals responsible for "that lovin' feeling" (adrenaline, dopamine, norepinephrine, phenylethyl-amine, and so on) dwindle. Suddenly your lover has faults. Why has he or she changed, you may wonder. Actually, your partner probably hasn't changed at all. You're just able to see him or her rationally now, rather than through the blinding hormones of infatuation and passionate love. At this stage, the relationship is either strong enough to endure, or it ends.

If the relationship can advance, then other chemicals kick in. Endorphins, for example, are still providing a sense of well-being and security. Additionally, oxytocin is still released when you're having sex, producing feelings of sat-isfaction and attachment. Vasopressin also continues to play a role in attachment.

Are We Alone in Love?

Only three percent of mammals (aside from the human species) form "family" relationships like we do. The prairie vole is one such animal. This vole mates for life and prefers spending time with its mate over spending time with any other voles. Voles even go to the extreme of avoiding voles of the opposite sex.

When they have offspring, the couple works together to care for them. They spend hours grooming each other and just hanging out together. Studies have been done to try to determine the chemical makeup that might explain why the prairie vole forms this lifelong, monogamous relationship when its very close relative, the montane vole, does not.

According to studies by Larry Young, a social attachment researcher at Emory University, when the prairie vole mates, the hormones oxytocin and vasopressin are released, as in humans. Because the prairie vole has the needed receptors in its brain for these hormones in the regions responsible for reward and reinforcement, the vole forms a bond with its mate. The bond is for that particular vole based on its smell—sort of like an imprint.

As further reinforcement, dopamine is also released in the brain's reward center when the voles have sex, making the experience enjoyable and ensuring that they want to do it again. And because of the oxytocin and vasopressin, they want to have sex with the same vole.

Because the montane vole does not have receptors for oxytocin or vasopressin in its brain, those chemicals have no effect, and the montane voles continue with their one-night stands. Other than those receptors, the two vole species are almost entirely the same in their physical makeup.

LOVE POTION NO. 9: TOP FIVE LOVE CHEMICALS IN THE BRAIN

T hose crazy behaviors people exhibit when in love are often chalked up to fate, but science attributes them to a cocktail of chemicals buzzing around our brains. When love overwhelms our actions, which five neurochemicals are largely to blame?

Love Potion No. 5: Testosterone

Although it certainly isn't the most romantic hormone pumping through our veins, testosterone—the compound responsible for facial hair and fast driving—is necessary to warm up our internal love engines. And it isn't just for the guys, either. Testosterone also exists in the female body to stoke physical attraction and sexual arousal.

Love Potion No. 4: Serotonin

Serotonin is a somewhat counterintuitive hormone to make the list, since it actually promotes feelings of calm and contentedness. But it's possible to chart the lifespan of a romantic relationship by tracing the roller-coaster ride of serotonin in the brain. During the early attachment phase of love, serotonin takes a backseat, residing at low levels, while other reward-regulating chemicals take over.

As a result of that serotonin dampening, people become borderline obsessed with their beloved, unable to focus or eat whenever apart. Eventually, once a relationship solidifies, the raphe nucleus in the brain stem begins to cook up more serotonin, eliciting those warm and fuzzy feelings of togetherness that typify longer-term attachment. The only downside of that serotonin upshot is the loss of excitement, also colloquially known as the end of the "honeymoon phase."

Love Potion No. 3: Oxytocin

What happens in the brain during an orgasm? One word: oxytocin. Sexual intercourse promotes feelings of attachment over time in large part thanks to oxytocin, which is produced in the ventral tegmental area of the brain. This chemical could aptly be nicknamed the "glue chemical" since it does such a swell job of binding people together.

As its levels crest, oxytocin calms and combats the early-phase intensity of romantic attachment, easing us into more stable relationships. Research confirms that in women especially, oxytocin fosters trust, happiness, and bonding. This is the same nonapeptide compound that bathes a new mother's brain, establishing the maternal link between her and child, and that spikes when we look at a picture of a loved one.

Love Potion No. 2: Vasopressin

Prairie voles, as we've learned, are often regarded as the animal kingdom's mascots of monogamy. The bonding oxytocin and vasopressin are the key neurological ingredients to the voles' faithfulness. Pair-bond-promoting vasopressin saturates the nucleus accumbens, a brain structure pivotal in sensing satisfaction. In doing so, this chemical drives vole pairs to lifelong coupling. In one experiment at Emory University, scientists blocked vasopressin receptors in prairie vole brains, resulting in an outbreak of adultery. Conversely, cranking up the levels of vasopressin in the brains of meadow voles—some of the prairie voles' promiscuous cousins—sparked a monogamy movement.

Love Potion No. 1: Dopamine

Once testosterone gets the job done of attracting two people to each other and igniting their sexual energy, pleasure-inducing dopamine is released during sexual intercourse. By powering the brain's limbic reward system, this chemical cultivates the addictive satisfaction that comes with the euphoria derived from eating a delectable meal or making love or even using cocaine.

That neurological treat dopamine doles out in the early stages of romantic attraction is partially responsible for the thrill we get when seeing the object of our affection and the craving to be with him or her in between. Indeed, studies have demonstrated that the brain, thanks to dopamine and its neurochemical cohorts, processes and manifests love much like an addiction, compelling people to settle down and reproduce.

Though high levels of dopamine are characteristic of early-stage romance and obsession, its presence is also a hallmark of satisfying, lasting relationships years later. Functional MRI scans of couples married at least twenty-one years who reported being madly in love showed above-average activity in the brain's dopamine factory, the ventral tegmental area. So while oxytocin and vasopressin are important for establishing monogamy, dopamine ensures that Pablo Neruda's "endless simplicity of tenderness" remains long after the honeymoon phase fades.

WOOING ON THE WEB: HOW ONLINE DATING WORKS

O ne of the basic human impulses is to fall in love. But a lot of obstacles can keep one from meeting the love of his or her life in today's world. Perhaps you hate the bar scene. Maybe dating coworkers is against company policy. Or you simply might not be in the right mood to meet your soul mate while trekking through the grocery store after a long day.

People of all ages, lifestyles, and locations have faced this problem for decades. In the last ten years or so, a new solution has arrived to help lonely hearts find their soul mates: online dating.

Online dating has its advantages and disadvantages. The variety of dating sites is constantly growing, with many sites focused on very specific groups or interests. In this section, we'll focus on the most basic type of dating site—one that works to bring two people together for a romantic relationship. Read on to learn what online dating is like, get tips for how to navigate it successfully, and find out how (and if) it works.

Online Dating: Creating a Profile

When you first arrive at an online dating site, you can browse through profiles without entering any information about yourself. The amount of information you can see about each user depends on the site. Some sites allow users to restrict access to their profiles to paying members. Photos might not be displayed unless you have a paid membership. This helps preserve anonymity, since a coworker or family member can't accidentally stumble across your profile. They'd have to pay for a membership to see a picture of the person they're reading about.

If you browse through a typical dating site, you will see hundreds of ads from people who are "looking for Mr. Right." Nearly everyone "enjoys a night out on the town, but also likes a quiet evening at home." It would be difficult to find someone who doesn't like a good sense of humor in a date.

So if you want to create a profile that stands out from the crowd, begin with the subject. Inject some humor into your subject line or include one of your interests. "Bogart fan seeking unusual suspects." "Come sail away with this boating enthusiast/Styx fan." This is the first thing people will see, and it needs to stand out from the crowd.

ONLINE DATING 101

Tip: Make sure you fill out the whole profile. Take your time and put some thought into it. Describing yourself may seem tedious or difficult, but if you leave sections blank or put in short, generic answers, you'll look like you aren't really interested. Avoid phrases like, "I wouldn't normally use one of these dating services, but my friends put me up to this." Remember, your target audience is other people who are using this dating service. You don't want to start off by insulting them.

Think of specific aspects of your personality that you want to highlight. Then, don't just state them—demonstrate them. Instead of, "I enjoy Stanley Kubrick films," say, "The other night I was watching *A Clockwork Orange*, and I found myself thinking it would be a lot more fun to watch and discuss it with someone else." Humor is especially important. Not everyone shares the same sense of humor, so saying "I'm a funny person" isn't sufficient. "I love quoting lines from Monty Python sketches and *Simpsons* episodes" gives other users a better grasp of your personality.

Another key to success is knowing what you want and putting it in your profile. That way, you'll get more responses from people who are looking for the same thing you are, whether you want to settle down with a long-term relationship or just want a date for Friday night. "I think there is more of a mental connection first by online dating," said one user, a teacher from New York. "Also, you know what you're looking for, not what your friends think would be 'perfect' for you."

To Be...or Not Meant to Be: The Science of Online Matchmaking

Once you've filled out a profile, online dating sites will provide a list of matches—people they think you are compatible with. How do they decide who matches up with who?

Sometimes, the process is very simple. Each profile has a list of attributes or interests that members check off. The more matching attributes that two profiles have, the higher "match percentage" the site will assign to them. Some sites, like Match.com, allow users to specify how important each factor is and assign it a different weight depending on how important it is to them.

For example, if you prefer blonds, but really have nothing against brunettes and redheads, then you can rank that attribute very low. If it's very important to you that your date has a college degree, you can rank that very high. Then the site will match you with a highly educated brunette sooner than a blond who didn't finish high school.

Some sites use very complex personality surveys and mathematical algorithms to match partners. Online matchmaking site eHarmony.com uses "twenty-nine key dimensions that help predict compatibility and the potential for relationship success." Their system was developed by theologian and psychologist Neil Clark Warren, who studied thousands of marriages to develop his "predictive model of compatibility."

Do such scientific methods work? Obviously, the dating sites claim they do. However, scientific personality tests completed with the guidance of a trained researcher do not have 100 percent accuracy. (It's closer to 75 percent.) And when you're sitting alone in your living room filling out a personality profile on a website, there is an even greater chance that the resulting matches will not be perfect. When you multiply the chance for inaccuracy by the number of users on a given dating site, complicated matching systems are probably not working much better than basic attribute-and-interest matching.

Fortunately, the main advantage of online dating is that it gives each user control over who they contact and with whom they subsequently communicate. It might take more work than relying on the site's matching system, but browsing through profiles yourself may ultimately be the best way to find the right person.

Specific facts and figures for online dating are hard to come by. For obvious reasons, each individual site tends to inflate membership numbers and success rates in its promotional materials. There are close to 100 million single adults in the United States alone. Of those, 40 million use online dating services. FriendFinder.com claims more than 11 million members. Eharmony.com claims responsibility for tens of thousands of marriages. But it's difficult to know for sure if either of these statistics is factually correct and if online dating actually drives more people toward long-term relationships than traditional matchmaking methods.

While some of the numbers may be fuzzy, one thing is certain—the use of online dating services continues in huge numbers. The online dating industry is predicted to reach $1.2 billion in revenue this year, according to *USA Today*, and the explosive growth of online dating mobile apps, such as SinglesAroundMe and others, will most likely continue to spur the use of online dating on the go.

ONLINE VS. OLD-FASHIONED MATCHMAKERS

A matchmaker is someone who personally interviews singles and pairs them off for dates based on his or her own judgment as to who would make a good match.

After each date, the singles give the matchmaker feedback on the compatibility and appropriateness of the match. The matchmaker uses this information to further refine his or her selections.

This differs from online dating sites mostly because the sites use a computer program to suggest potential matches, and that computer program doesn't adjust its thinking based on your feedback. Ultimately, it is up to the user to choose whom to contact or go out on a date with. With a matchmaker, you're leaving the decision in the hands of another person.

Another important difference is cost. Matchmaking services can cost thousands of dollars, while typical dating-site fees average between twenty and thirty dollars per month. Matchmaking services have an obvious appeal for those who want a more personal touch, but for the cost-conscious single, dating websites may be the better choice.

AGENTS OF AMOUR: HOW MATCHMAKERS WORK

Matchmakers have paired up couples for centuries, and laxer social rules and online dating don't seem to have slackened their market. From traditional *shadchans* to professional Cupids, what does the modern matchmaker do?

In Orthodox Jewish communities, matchmakers have paired up couples for centuries, but their niche profession entered the American pop-culture lexicon in 1964. That year, the musical *Fiddler on the Roof* debuted on Broadway, scoring nine Tony Awards before being adapted into a 1971 Academy Award-winning film. Among the popular show tunes included in the production is the wistful "Matchmaker, Matchmaker," featuring a girl named Tzeitel who is pining for a handsome, romantic suitor, rather than a dispassionate arrangement. Cultural tradition would have dictated that a local matchmaker named Yenta find a prospective husband for Tzeitel, rather than allowing fate—and sexual chemistry—to take its course.

Set in the early 1900s, *Fiddler on the Roof* takes place

during a time when marrying for love was still a relatively new phenomenon. Until the Enlightenment and the Industrial Revolution in the West, marriage was widely perceived as an economic tool rather than an amorous union. From the nobility on down to the lower classes, parents strategically fixed up their children to secure or expand property, reap wealth from bridal fees called dowries, continue blue bloodlines, and in families rich in daughters and short on cash, to alleviate the financial burden of having to feed and clothe women who couldn't strike out on their own.

Historically, marriage in which two people meet serendipitously and get hitched is very much the exception to the rule. Although Yenta the matchmaker in *Fiddler on the Roof* is portrayed as the enemy of Tzeitel's girlish longing, her role in scheduling nuptial destiny was far more common than an audience might think.

Even today, roughly 60 percent of marriages around the world are arranged, often mediated by families as the result of cultural or religious custom. In laxer American society, setups are still routine. Just under 30 percent of heterosexual couples were first introduced by mutual friends playing Cupid. An increasing number of people also have turned to the Internet to seek out long-term relationships, and as of 2009, online dating sites that serve as virtual matchmakers were the second-most common way for couples to meet. In fact, 61 percent of same-sex couples found each other online.

But when all else fails—or if there simply aren't enough hours in the day to hunt for a soul mate—people can still call up their nearest Yenta and get a professional matchmaker on the case.

Modern Matchmaking and the Business of Love

You might think real-world matchmakers would be threatened by the explosion of online dating sites, but interestingly, they're not. In fact, today's professional matchmakers attribute much of their face-to-face businesses' success to the popularity and attendant drawbacks of online dating. Many anecdotes from men and women who have turned to professional matchmakers include online dating horror stories that compelled them to place their love lives in more responsible hands than the Internet.

The Matchmaking Institute, which trains people in the fine art of fixing up, estimated that there were 1,500 such professionals in the United States as of 2006, and that number has likely risen since then. The $250-million industry is dominated by women who may have been formally trained through the Matchmaking Institute, parlayed expansive social networks into a profitable service, or inherited the itch to teach adults how to date from a familial interest in matchmaking.

One of the most famous matchmakers in the United States and star of the reality television series *Millionaire Matchmaker* is Patti Stanger. She started getting couples together in the seventh grade. Her early start isn't terribly surprising, considering that both her Jewish mother and grandmother had established themselves as local matchmakers in her New Jersey hometown. Meanwhile, Janis Spindel, founder of Manhattan-based Serious Matchmaking, Inc., ditched a career in the fashion industry after reportedly pairing up fourteen serious couples in a single year. Both claim impressive results: Stanger reports a 99 percent success rate, and Spindel says she has sealed the deal on more than 900 couples since 1993.

Name-brand professional matchmakers like Stanger and Spindel also do well for themselves financially. Unlike online dating sites that are free to join or charge nominal fees, professional matchmakers don't cater to a frugal crowd. Stanger's Millionaire's Club, for instance, costs $40,000 for a yearlong membership, and clients can fork over up to $200,000 for more personalized services and individualized attention.

For a more representative price point, about a third of matchmaker clients spend $3,000 to $5,000 per year each on the dating services. In New York, the top market for matchmaking, the average professional earns $78,000 per year, which is probably well below the income level of the typical person willing to pay a little extra for love—or at least a lot of dates.

MATCHMAKER HOTSPOTS

Where are people most likely to hire professional matchmakers in the United States? The following top markets, according to the Matchmaking Institute, may surprise you:

+ New York
+ Los Angeles
+ Chicago
+ Atlanta
+ Minneapolis-St. Paul

The Professional Matchmaking Process

Professional matchmaking was once a service used almost exclusively by wealthy men with the disposable income to have someone else sort through the choppy waters of the dating pool on their behalf. Many matchmakers, including New York's Janis Spindel, work exclusively with male clients, and the standard business model was largely built on the premise of bringing potential brides to rich, single men.

But that guy-seeking-girl tide has turned, and just as many—if not more—gainfully employed women have begun turning to matchmakers to make their romantic dreams come true. Industry statistics report that women typically comprise 60 percent of matchmaker customers, in fact.

Matchmakers attract clients in one of two ways. Either people seek them out through advertisements, online searches, or word of mouth, or the matchmakers proactively recruit wedding-band-free singles at parties, high-end restaurants, airports, and other places where affluent adults congregate. Before potential clients purchase an official membership, matchmakers often conduct an initial consultation, sometimes for a nonrefundable fee, to find out what type of relationship the client is interested in and with what type of person.

From there, the official process begins. Membership dues are paid, or the matchmaker may refer the person to another service better tailored to his or her interests. On the other side of the set-up equation, men and women who wish to be included in a matchmaker's bank of potential date picks for clients may come directly from the professional's social network, or they may attend formal recruiting sessions or auditions. Those ladies- and gentlemen-in-waiting may also fork over a fee to get a spot on one of those matchmaker lists.

To determine the best possible coupling, a matchmaker

will first dive into a client's romantic psyche by finding out information, including:

- ♥ Family background.
- ♥ Educational background.
- ♥ Hobbies and interests.
- ♥ Religious background and observances.
- ♥ Personal values and morals.
- ♥ How many children he or she wants (if any).
- ♥ Previous relationships and relationship deal breakers.

As matchmakers sift through their Rolodexes to find ideal suitors, clients may undergo more prep work to prepare themselves for the dating process. Particularly with higher-end services, matchmakers will double as dating coaches, teaching clients how to spark conversation and avoid dicey topics, as well as weaning them away from negative inter-personal habits, such as talking about themselves excessively rather than focusing on the other person. Makeovers may also be in order, and image consultants may assist clients with sprucing up their wardrobes, addressing unflattering hair-styles, and sculpting their bodies.

Getting clients into the dating mix will vary, depending on the matchmaker. Some serve as escorts to parties and introduce clients around to appropriate singles; others may arrange events specifically for clients to meet a number of men or women on their dating rosters. Or, if a matchmaker has somebody in mind who seems like a good fit for the man or woman in question, a one-on-one date may be arranged.

After a first date, the matchmaker will check with each party to find out how things went from both perspectives. That information allows the matchmaker to gauge whether a client needs more date coaching or if the match can move forward. The best-case scenario is for an arranged couple to hit things off and eventually head down the aisle, but those looking for love shouldn't expect things to happen over-night. Generally, these pricey interventions last for at least a year, which is far longer than matchmakers in other cultures expect couples to get to know each other before making a lifelong commitment.

WHAT YOU SEE IS WHAT YOU GET

More often than not, matchmakers won't show clients photos of dates before face-to-face meet-ups. According to a 2009 survey conducted by the Matchmaking Institute, only a third of professionals give clients photographic sneak peeks of who might sweep them off their feet.

⟩ Religious and Cultural Matchmaking

Religious faith has long held a strong link to matchmaking and arranged marriage. Per Jewish tradition, God was the original matchmaker, creating Eve out of Adam's rib so that the two could share company and procreate. Documented in the Bible, the Torah, and the Talmud, matchmakers have held a position in Orthodox Jewish society throughout its history and the Jewish diaspora.

Fathers customarily bore the responsibility of selecting adequate grooms for their daughters and might request assistance from a local shadchan to seek out an eligible bachelor. Matchmakers might then team up with rabbis to pair young men and women in the community, something that still takes place in orthodox communities.

The Torah dictates payment to a shadchan, but that doesn't always happen. Some Jewish matchmakers refuse to accept any remuneration, considering matchmaking a divine calling that they pursue as a form of charity.

Similar to secular professional matchmakers, Jewish shadchans might inquire around to find out about a young man's character, personality, religious observance, family, and professional prospects before proceeding with the fix-up. Jewish matchmaking focuses more on shared family background and kindred morals than romantic attraction, and, likewise, the relationship-building is reserved for the postnuptial years. For that reason, once the preordained couple meets, they aren't expected to carry out an extended courtship. The young man may pop the question after only a couple months, if not sooner.

In Southeast Asia, arranged marriage remains a common custom, and the family often functions as matchmaker. With marriage a cornerstone establishment of the Hindu faith, the matchmaking tradition has existed in India, for instance, since the fourth century. Even in the twenty-first century, about 90 percent of Indian marriages are set up. Boys' families are generally the ones that initiate a search for a bride and may also solicit a matchmaker to ensure that a girl's family line and astrological signs are compatible.

Younger, more urban generations have sought more autonomy in their romantic lives, but even in the United States, some Indian singles continue to keep the family involved in their marital decisions, allowing them to vet or even choose potential suitors.

Marriage can also be a group effort within Muslim communities around the world. Older women nicknamed "aunties" and family members may play matchmaker by identifying potential mates in their social networks. Afterward, relatives arrange chaperoned meet-ups between the possible bride and groom. If it isn't a match, however, either party is permitted to give a thumbs-down on the marriage-to-be, in accordance with stipulations in the Koran.

Though match-made marriage might seem archaic by Western standards, statistically, these marriages are far more successful than love-made unions. The United States decries its abysmal divorce rate hovering around 50 percent, but only a slim minority of 5 to 7 percent of arranged marriages dissolve. The low number could be partially attributed to cultural, religious, or legal restrictions against divorce. But that stark gap has led some to wonder whether people are all that good at picking partners, or whether outside parties can better spot a long-term spouse.

WHAT IN THE WORLD IS A "YENTA"?

Perhaps due to the matchmaker character Yenta in the popular musical *Fiddler on the Roof*, the Yiddish word "yenta" is sometimes—and incorrectly—used interchangeably with "matchmaker." The correct Yiddish term for a matchmaker is a "shadchan" for a man and "shadchanit" for a woman. In fact, "yenta" merely refers to an old woman with a penchant for gossip, much like the fictional Yenta, who also happens to be a shadchanit.

⇉ The Science of Matchmaking

Whether matchmakers play Cupid as a profession to make money, an adherence to religious doctrine, or a following of cultural custom, they often use the same ingredient to spot Mr. or Ms. Right: common background. Generally, matchmakers seek out people who come from similar socio-economic strata, nearby geographic locations, identical education levels, and so forth. These unromantic basics are the primary determinants of whether a matchmaker foresees a fit, and for good reason.

Although "opposites attract" has become a well-trod trope, and unlikely pairings seem to make for the most enchanting stories, those are the exceptions to the rules of human mating. The tried-and-true bond of long-term relationships isn't a fleeting sexual fizzle but mutual upbringings and experience, or assortative mating in academic-speak. Mutual attraction and interpersonal chemistry are merely the set dressings that transform a platonic relationship into a loving one for the long haul.

The Westernized notion of marrying for love and passion might actually be an example of the blind leading the blind. Caught up in the dizzying sparkle of the moment, people might not realize that the foundational aspects of long-term relationships are missing and thus take time to look before they leap over the threshold into marriage. And while the lovelorn might have an ideal list of qualities they believe add up to their perfect partner, research suggests that those must-haves might be off base.

A study published in 2008 in the journal *Evolutionary Psychology* highlighted a disconnection between the types of partners participants idealized and the specific qualities they sought out. While men and women described their dream dates as having a similar personality to themselves, the components of what they were looking for—conscientiousness, extroversion, stability—were often more complementary, leading the researchers to conclude that people may lack self-awareness in understanding the type of person who would best suit their needs.

On the other hand, romantics can take heart from a 2012 analysis of online matchmaking published in the journal *Psychological Science*. Virtual matchmakers pair up profiles on the basis of complementary personality traits and interests, but lead author and Northwestern University psychologist Eli Finkel found the algorithms lacking. While both online and real-life matchmakers can follow the rules of assortative mating, they nevertheless miss out on the spice that distinguishes friends from lovers: relationship aptitude. In short, relationship aptitude denotes the combination of life factors, personality, and interests that ignites a unique chemistry and desire for intimate companionship between two people. It's the secret ingredient in the otherwise predictable recipe for compatibility.

Professional matchmakers are trained to predict whether people will discover relationship aptitude, but the result can be hit or miss, of course. By helping clients, family, or friends to sharpen their sights toward potential partners who have those foundational assortative mating criteria, however, matchmakers can very well prod the willing toward a happily-ever-after.

RAPID ROMANCE: HOW SPEED DATING WORKS

You've probably heard of speed dating by now. It has so permeated Western popular culture that even those who aren't looking for love know what it is. The concept was introduced to the American dating scene at the turn of the twentieth century by Rabbi Yaacov Deyo and his wife, Sue, who founded their own service, SpeedDating, in Los Angeles. The service was based on an old Jewish tradition: helping young, single Jews meet others in the faith. This tradition of creating a *shidduch*, or a match, called for Jewish singles to be kept in the dark about each other until the time for matchmaking came.

Today, modern speed dating is still rooted in shidduch, but with formal dating services replacing the role of the rabbi and his wife as matchmakers. These services compile the data from brief encounters between daters and then inform each attendee of the results, allowing interested parties that scored a "match" to pursue another meeting with each other.

So what is speed dating, and more specifically, how does speed dating work? Perhaps even more importantly, *does* speed dating work? Read on to learn more about this old Jewish tradition that has become an international phenomenon.

The Rules of Attraction

Speed-dating events are most often held in restaurants and bars, although events are cropping up in other places, like student centers on college campuses. Participants are asked to register ahead of time to ensure an even ratio between men and women, although some services now offer registration at the door.

Inside the venue, speed daters will find that tables are arranged to accommodate two participants at a time. One set of the speed daters, usually women, stay seated at the same table, and the opposite group moves from table to table. This table-hopping method has been compared to musical chairs. The difference is that when the bell rings or buzzer sounds, the next seat the dater takes is predetermined. The speed dater progresses from table to table until each participant has had a chance to meet the other.

Depending on the company, a speed date may last from three to eight minutes, although some go as long as ten minutes. At the end of the date, each dater makes a note if he or she would like to see the other person again.

The number of dates held in an evening can vary, but

most services hold ten or fewer. In less than two hours, each person has ten chances to meet the love of his or her life.

After the event, the speed daters turn in their date cards to event organizers. They may be contacted via e-mail the following day or asked to log on to a website to enter the names and ID numbers of people whom they would like to see again. If two speed daters have registered a mutual interest in seeing each other again, the pair receives each other's contact information. From there the couple can contact each other to arrange another meeting or date.

8 Minute Dating, a speed-dating service based in Boston, maintains the policy that none of their speed daters are allowed to ask for anyone else's last name or phone number. Other services ask speed daters not to discuss what they do for a living or where they live. The idea is for the couple to pursue a connection based on mutual attraction rather than one person doggedly pursuing the other.

Companies like 8 Minute Dating can hold different speed-dating events scheduled at the same time in different cities by franchising their services. Events are put on by local organizers on behalf of the company. To this end, most websites for speed-dating services have a page dedicated to becoming an event organizer.

TOP TIPS FOR SPEED DATING

+ To attract each other, men should wear blue and women should wear red.

+ Women should wear the scents of vanilla and cinnamon to attract men.

+ Men can increase their chances by sporting a black licorice scent.

+ Refrain from getting drunk. (Trust us, your potential mates don't want to entertain the possibility of a drunken date.)

+ Smile. It goes a long way!

⋧ Specialized Speed Dating

While the concept of speed dating itself is already tailored to help busy people meet the right person, it can be even further customized. Most speed-dating services offer specialized speed-dating events. Some match tall men to petite women; others are held specifically for Jews, Christians, Muslims, or members of other faiths. Still others are designed for people who share a love of Broadway plays and musicals, differently abled singles, iPad junkies, single parents, and even millionaires.

By hosting events for people who already share at least one

thing in common, organizers can assure a better chance for a match. For example, a "single athletes" event may be more likely to produce a connection between two people based on their shared enjoyment of a particular sport.

Just about all speed-dating events also have age restrictions. Some are wide ranging, and some only require that speed daters be twenty-one or older to meet the age restriction of the bar where the event is held. Others may require that participants fall within the ages of twenty-five to thirty-five for one session, and forty to fifty-four for another session later the same evening, for example.

Whatever the age, the question still remains: does speed dating work? Could be. Science has been busy uncovering facts about love and attraction that may support the concept of finding love in ten minutes or less.

YOU SMELL LIKE...LOVE!

The ABC News show *20/20* sent pairs of twins out on speed dates to test whether pheromones (scent molecules humans emit that are believed to attract other humans) might play a role in attraction. Twins were given a fragrance to wear, one with added pheromones and one without. It was a blind test, so neither twin knew who had been doused in pheromones.

The results were positive in favor of the molecules' effectiveness. One female who wore pheromones had nine out of ten guys say they wanted to see her again. Her identical twin attracted only five of the same men. Not necessarily conclusive, but enough so that some speed-dating websites now offer perfumes laden with pheromones—just in case.

Speed Demons: Does Speed Dating Really Work?

Studies suggest that speed dating should, in theory, work. If we can determine whether another person is a good match for us in just a minute or two, then speed dating is an optimal approach to selecting a mate. Why waste time on some jerk when you've already decided that you'll most likely never speak to him again? Speed dating also offers a structure that—in its brevity—encourages polite behavior. And with the speed-dating service ringing a buzzer that signals the end of a couples' time together, participants can relax knowing that they can largely avoid any awkward end-of-date moments.

But do the rules of attraction still apply in speed dating? University of Pennsylvania psychologists examined more than 10,000 client responses from HurryDate's database and found that in the context of a speed date, the usual rules of

attraction go out the window. Factors like religious affilia-
tion and earning potential—usually viewed as very important in dating—are replaced by behavioral cues. These cues
provide the basis of attraction in a setting where time is of
the essence and split-second decisions are made.

The University of Pennsylvania researchers determined that HurryDate's three-minute format was longer
than necessary—three seconds is about all it takes, said
one researcher.

Another study conducted by Stanford, Harvard, and
Columbia University researchers also found that women in
the speed-date setting throw out traditional requisites for a
mate, like intelligence and sincerity, and go instead for physical attractiveness. So, too, do men, but this represents no
change, as men traditionally report physical attractiveness at
the top of their list of desirable qualities in a mate.

This same study also found that the smaller the pool of
potential candidates, the more likely women were to want
to see any of the given men. As the number of men at the
speed date simulation increased, the number of men the
women wanted to see decreased.

Scientific study has come up with quantitative evidence
that speed dating can work in the selection of a mate. But
there's also plenty of qualitative evidence suggesting that
speed dating can fail. Some speed daters report the scene

is fraught with sleazy and insincere individuals. Others find the candidates somewhat lacking. "I've never seen so much desperation in one area," reported one college student who tried speed dating at an event at the University of Buffalo. While comprised of brief, timed encounters, speed dating also generally requires that the dater sit through the entirety of a date, with little or no chance of the escape usually found in other, less structured settings.

In stark contrast to these objections, speed-dating service websites are brimming with anecdotal evidence that the technique works. Most sites are long on success stories and display wedding photos of couples who met through their service's events. 8 Minute Dating boasts that 62 percent of its clients find a mutual interest with another speed dater. "Compare that to the bar scene!" the website dares.

SPEED DATING IN POPULAR CULTURE

Speed dating is far too fraught with potential pitfalls and is too obvious a reflection of the fast-paced society in which we live for art not to emulate it. Some of the places speed dating has appeared are:

+ *Speed Dating* (2010)—Three guys try to save a failing nightclub and meet women by introducing speed dating.

+ *Speed Dating* (2007)—The Irish film chronicles a man who finds he has become addicted to speed dating.

+ *The 40-Year-Old Virgin* (2005)—Steve Carrell's character is talked into trying speed dating. He has little success.

+ *Hitch* (2005)—Eva Mendes gives speed dating a shot after Will Smith turns out to be a jerk.

+ *8 Minutes To Love* (2004)—Sandra Oh finds her boyfriend engaged in a speed-dating event. She gives him a speed date to win her back.

+ *How I Met Your Mother* (2005)—In one episode, the stars of this show hosted a speed-dating event at Grand Central Station in New York City.

+ *The Graham Norton Show*—The late-night British talk-show host sent his cameras to a London speed-dating event on one episode.

+ *Speed Dating, the Musical*—This play premiered in 2008.

MÉNAGE À TROIS...
OR MORE: HOW
POLYAMORY WORKS

L ove is often described as two halves coming
together to form a whole. But what happens when
you have more than two halves?

Romantic comedies and love songs tell us that we'll
find the person who will make us complete, and then we'll
marry him or her, have children, and grow old together.
But the idea of marrying our soul mate is a relatively new
one. For many centuries, people married someone their parents deemed fit, and then they pursued love with others, no
questions asked. Some people claim that rising divorce rates
and high incidence of infidelity are proof that monogamy,
even with someone you truly love, just doesn't work.

So where does that leave us? Could monogamy be a bad
system? What if it takes more than one person to make you
feel complete? After all, we place rather tall orders with our
soul mates—we expect them to like the same types of movies,
be compatible sexually, and have the right words to say to us
no matter what happens. Some people would argue that a
single person can't fulfill all those needs, and that it's foolish to

make one person try. These people practice polyamory, the practice of having multiple romantic relationships. But they claim they're not cheating or running around. Rather, a central tenet of polyamory is garnering your partner's consent to date and fall in love with multiple people.

It can be hard to wrap your head around polyamory, if only because monogamy is set as the default for our society. So first, here is what polyamory is *not*.

♥ It's not about sex with a bunch of random people. While polyamorists certainly do have sex with multiple partners, they usually also have emotional relationships with them.

♥ It's distinct from polygamy, which we tend to associate with fundamentalist Mormons who practice plural marriage. In those communities, men marry multiple women, while in polyamory, both genders have the opportunity to explore connections with other people.

So now that we've established a little bit about what's not polyamory, let's take a look at what polyamory actually is. Is it really possible to love more than one person? Don't people get jealous? And if we hardly have time to maintain one good relationship in today's busy world, how do people find time to manage three or four?

"I Do...and I Do Again"? Examples of Polyamorous Relationships

It's impossible to know how many people practice polyamory, because most forms ask for things like a spouse's name, leaving no space for people to write in an additional boyfriend or girlfriend's name. However, awareness of polyamory has grown tremendously because of the Internet, and according to current estimates, based on web usage and online polls, as much as 10 percent of the U.S. population self-identifies as polyamorous.

But who are these people? What do we know about polyamorists? According to a 2002 survey conducted by the polyamory awareness site Loving More, 40 percent of polyamorous people had a graduate degree (compared to 8 percent of the general population) and 30 percent identified themselves as pagan. Other anecdotal data suggests that most polyamorous people are white and in their thirties, forties, and fifties. In addition, many people who practice polyamory identify themselves as bisexual.

There is no one way to practice polyamory, but let's consider a few hypothetical set-ups. Let's say Ann and Bob are a married couple who practice polyamory. While Ann and Bob live together with their one child, Ann has a boyfriend who lives fifteen minutes away that she stays with two nights a week. That boyfriend, in turn, has another girlfriend who is friends with Ann.

Bob has a girlfriend that he stays with one night a week, as well as a boyfriend who lives out of town that he sees occasionally. Ann and Bob have met each other's partners and frequently host dinners where all of the significant others come over to socialize. Ann and Bob's child is on good terms with all of the partners, but doesn't realize that they are his mom and dad's boyfriends and girlfriends.

Here's another example: Ann and Bob are a married couple, and they form a quad with another couple named Cathy and Dave. Ann frequently goes out on dates with Dave, and they spend a night together in a hotel once a week. Bob and Cathy do the same. The four of them are considering buying a home together, and none of them want children. They consider themselves polyfidelitous, meaning they're not open to other partners outside of their group right now.

One last example: Ann and Bob are married, but they're both in love with a woman named Cathy, who just moved in with them. Each person has his or her own bedroom, but depending on how they're feeling, two of them may spend the night together—or all three may spend the night in a king-size bed. Both Ann and Cathy would eventually like to carry a baby fathered by Bob, though they plan to raise the children all together. They have no other partners at this time but would okay with any of the three finding someone outside the group to date.

Again, these are just examples of how some polyamorous situations might go. Sometimes, keeping up with all of the partners involved can take a massive organizational chart, but sometimes it may be as simple as a triangle (three people in love with each other equally). But how do people make this work?

Dating Doubles: The Logistics of Polyamory

Polyamory involves a lot of talking—so much so that "communicate, communicate, communicate" is considered one of the core tenets of polyamory. Though polyamory may seem like a bit of a free-for-all, it can actually involve a lot of ground rules. Remember, everyone has to know what everyone else is up to in carrying on outside relationships. That involves a lot of conversation (as well as, possibly, a conversation about how much detail you want about what your partner is doing with other people).

It is also crucial to negotiate boundaries to ensure that each relationship receives ample time. For example, a woman may request that her husband only spend three nights a week with his other girlfriend. New partners usually have to meet with already-existing partners and get their approval (or at least avoid a veto; the power to nix a potential partner is usually the right of someone already in the relationship). All couples face questions of where to live and how to allocate resources, but talks get more intense with so many players involved.

Sex also comes with a lot of guidelines, so that everyone avoids sexually transmitted diseases. A married couple, for example, may be body fluid monogamous—meaning that they exchange body fluids without the protection of a condom—but they may have a rule that a condom must be used in encounters with other sexual partners. There may also be rules about how often someone must be tested for STDs in order to remain in the relationship.

Along with communication skills, good scheduling abilities are essential to the polyamoric lifestyle. Shared online calendars, such as the one provided by Google, can be vital to remembering which girlfriend has a work event and which needs to be at her son's school. It might be disappointing for someone if their boyfriend can't come to dinner on a night when he's already scheduled to be with another partner, but again, talking about these kinds of issues and feelings is expected—especially when the feeling at hand is jealousy.

Jealousy, worry, or insecurity about your standing with someone you love is a universal emotion, and the chances for it are multiplied infinitely when you know that your partner is sleeping with someone else. And even though polyamorists know what they're getting into, they're not exempt from experiencing the green-eyed monster. However, rather than use a feeling of jealousy to fly off the handle, polyamorists try to assess themselves and communicate with their partners to figure out what the true issue is, and how it can be resolved.

Does this seem like a lot of work? It can be. So why choose this lifestyle? Read on to explore some of the potential benefits.

Polyamory Pros

People who practice polyamory probably don't think monogamy is a realistic practice. It's inevitable, they might argue, that we'll have the urge to pursue or sleep with someone who is not our spouse or life partner. By recognizing this factor and working around it, they've found a way to keep important relationships intact. Rather than enduring a devastating breakup over a dalliance, they believe they've found a way to keep the person that they love in their life, even if he or she seeks to fulfill needs that they can't. In polyamorous lingo, there's even a word for feeling joy over the fact that your significant other has found happiness with another person: comparison.

Because their dating options aren't limited by saying "I do" or making a commitment to another person, polyamorous people often cite freedom of choice as a main motivator for their lifestyle. There is less pressure to find that perfect person to grow old with. Rather, polyamory allows a person to seek out an entire network of people that meet his or her emotional and physical needs, which allows for lots of different kinds of intimacy and support.

On the most practical level, that might mean being able to avoid watching an afternoon of football or ballet if a partner has another partner who enjoys that activity. Having such a wide array of relationship experiences might mean not becoming bored or complacent in any of the relationships, and it might allow individuals to get to know themselves better.

One study indicates that this kind of freedom and choice can strengthen relationships, not hurt them. According to an analysis published in the *Electronic Journal of Human Sexuality* in 2005, polyamorous couples who had been together for more than ten years said "love" and the "connection" were the most important factors in their longevity. Monogamous couples, on the other hand, often cite religion or family as the most important reasons for a long-term commitment.

And when polyamorous people raise children, they have several partners, not just one, to help with parental duties like driving to soccer practice and figuring out homework. Though no research has been completed on the long-term effects of growing up with polyamorous parents, early findings from a study at Georgia State University indicate that kids surrounded by multiple adults benefit from the wealth of resources that a polyamorous relationship can provide. Still, kids are often the reason that many people stay quiet about polyamory. We'll explore some drawbacks of polyamory next.

Drawbacks of Dating Many

Though polyamory's profile has risen courtesy of the Internet, it seems highly unlikely that the practice will ever become widespread. Some people simply can't fathom the lifestyle, and most governmental and legal systems around the world are set up to recognize the legal rights of a married man and woman heading a family. (Witness the difficulty gay couples have had in trying to get another form of family recognized in many countries.)

Also, because polyamory seems so outside the norm, the stigma of this kind of lifestyle keeps many people "in the closet," so to speak. Polyamorous people may not tell their coworkers, their friends, or even their parents about the number of people they've chosen to love, out of fear of personal repercussions. Women, in particular, are known to keep quiet about polyamory, due to social stigmas about women who sleep around.

Women who have kids have a particular need to worry about keeping their lifestyle a secret. In 1999, polyamory made the news when a young child was removed from the custody of her mother, April Divilbiss. Divilbiss had appeared on an MTV documentary about polyamory with her two boyfriends, neither of whom was the child's father. The child's paternal grandmother sued for custody and won. Even though court counselors filed reports that the child's home was safe and happy, the judge ruled that Divilbiss' lifestyle was immoral and depraved.

And of course, one drawback of polyamory is simply how complicated it can be to juggle so many relationships. While having more partners might bring more pleasure, it can also come with more problems, and breaking up with one person can have ramifications beyond just the two people who have ended their relationship. Still, there's no evidence that polyamorous relationships break up any more or any less frequently than monogamous relationships. In love, everyone takes the same chances.

THE HIERARCHY

People who practice polyamory sometimes refer to a primary partner, a secondary partner, or a tertiary partner. This form of hierarchy can be a helpful way to prioritize and schedule time with partners. A primary partner may be a spouse or the person who has been there the longest, and he or she may take up the bulk of a partner's time. Secondary partners may have their own primary partners and thus require a slightly lesser commitment. Some polyamorous people, however, frown upon a hierarchal arrangement and divide their time fairly equally among their partners.

HAPPILY EVER AFTER: HOW MARRIAGE WORKS

Saying "I do" is one of the most important experiences of a person's life, but like any significant life event, it can get complicated. Depending on where you live, there may be laws governing who can get married and how a marriage license can be obtained. Some couples also face the decision of whether or not to have a prenuptial agreement. And then there's that whole rising-divorce-rate factor. In this section, we'll take a look at the benefits and pitfalls of marriage, and peek into the science behind long-term love.

Interpersonal Ingredients: Hallmarks of Happy Couples

It doesn't take a trained statistician to guess certain key ingredients to a happy marriage. According to the 2011 report from the National Marriage Project, husbands and wives cited sexual satisfaction, commitment, and positive attitudes about child-rearing among their respective must-haves. Add to that solid communication skills and extramarital social support, and the basic toolbox for weathering the years together is filled.

But pursuing love throughout a marriage also involves a degree of selfishness, come to find out. Stony Brook University psychologist Arthur Aron emphasizes the importance of finding self-expansion in a marriage as a means to long-term satisfaction. Couples that stoke each other's senses of learning, adventure, and intrigue—either alone or side by side—are unwitting experts in self-expansion, seeking novel experiences that enrich themselves and, by extension, their relationships. In exchange for that self-expansion, satisfied partners statistically perform five acts of generosity for every one instance of bickering or whatnot.

If those are the mechanics of a long-term, enjoyable marriage, what does that lasting sweetness feel like? In short, it feels a lot like falling in love. In 2009, psychologists Bianca P. Acevedo and Arthur Aron at Stony Brook University debunked the common gloomy notion that romance eventually fades as a marriage ages. What ebbs over time, rather, is obsession.

As couples bond over the years, the passion doesn't necessarily dwindle, just the overwhelming obsession and anxiety that initially came with the budding relationship. Early-stage romantic love is characterized by a nagging obsession for one's beloved, so strong and blindsiding that it's considered a threat to the human metabolism, gobbling up an astonishing amount of focus and energy. In that light, successful marriages are akin to trading out honeymoon suites for penthouses.

JUMPING THE BROOM

"Jumping the broom" is a popular euphemism for getting married, much like the phrase "tying the knot." The phrase originates from a tradition that was common among African-American slaves. Because slaves were often prohibited from officially marrying, they had to come up with their own traditions and ceremonies. Couples would jump over a broom that was set on the floor, symbolizing the beginning of a new life together. The tradition is still used by some African-American couples today. It is believed to have originated from African customs, though ancient Celtic tribes had a similar practice.

At the same time, that isn't to say that making marriages work is a cinch, as evidenced by the fact that Americans' lifetime chance of divorce hovers between 40 and 50 percent. For all of these interpersonal tips for nuptial bliss, the difference between the couples who do and don't endure can be seen clearly in the brain.

Neurological Necessities: Long-Term Love in the Brain

From an evolutionary perspective, that mind-altering joy of romance ensures that humans not only will want to reproduce, but also stick around long enough to rear a child. For that reason, UCLA psychologist Martie Haselton, who studies our primal compulsions to mate, sterilely refers to love as a "commitment device" reinforced by natural selection to direct our attention toward hearth and home. In slightly sweeter terms, the human brain—thanks to natural selection—processes romance like a box of chocolate.

The limbic reward system that similarly springs to action when we receive cash bonuses or sip on a glass of bubbly reacts positively to wooing someone who tickles our fancy. Functional MRI technology has illuminated the specific areas of the brain aroused in people freshly in love, which explain the swirl of positive and negative emotions that wash over us when we're blissfully together and fretfully apart.

In 2010, a team of researchers compared that fMRI data to brain scans of people who had been married around twenty years and reported still being in love with their partners. Despite the multiple years since each couple first locked eyes, the brain activity among those experiencing head-over-heels new love and those who had been together for decades appeared remarkably similar.

Speaking of things appearing similar, apparently long-term relationships of at least twenty-five years can even make partners *look* like each other.

Back to the sexy brain chemistry of long-term love. When both sets of people were shown photographs of their respective partners, two key brain regions lit up excitedly in both sets of people: the ventral tegmental area (VTA) in the center of the brain and the nearby caudate nucleus. Those structures help drive motivation and decision-making in pursuit of reward. The prize of being with their beloved refreshes the pleasure-inducing dopamine supply in people's brains. In addition, those neurological regions are packed with receptors for oxytocin and vasopressin, neurochemicals that promote bonding and monogamy.

What didn't sparkle at the sight of the long-term lover were areas associated with anxiety and fear, such as the amygdala. As mentioned earlier, self-reported psychological surveys and assessments denote the absence of addled obsession among people who have been happily married for an extended period of time. Instead, brain sites that also promote maternal attachment lit up at the sight of a life partner, demonstrating how that obsessive early-stage love gives way to calmly passionate commitment. But aside from that maturational distinction, the primary chemicals that preserve long-term marriage—dopamine, oxytocin, and vasopressin—are the same ones that gave us goose bumps to begin with.

For the 40 percent or so of modern marriages ending in divorce, those neurotransmitters eventually wear out their welcome, and the brain's reward system develops a tolerance of sorts for respective partners—similar to how an addict might develop a tolerance for a certain amount of drugs over time. The sizzle fizzles, sapping the interpersonal spice that initially drew two people together. Yet from a more sentimental stance, that's also what makes long-term love such a treasured bond—it's rare to find and even rarer to preserve 'til death do you part.

MARRIAGE BY THE NUMBERS

The demographics of marriage have changed dramatically in the last half-century, with the median age for a first marriage increasing since the 1950s. Divorce and remarriage rates have also increased over the last fifty years. Of people born from 1935 to 1939, 21 percent of men and 51 percent of women were married by age twenty. Of people born from 1975 to 1979, 8 percent of men and 18 percent of women were married by age twenty. In 2003, the median marriage age was twenty-seven years old for men and twenty-five years old for women. Clearly, young people are choosing to wait longer than their parents before taking the plunge into marriage.

Benefits of Marriage

Besides love and companionship, marriage offers many benefits, especially in the eyes of the law. In fact, 1,138 federal benefits, rights, and responsibilities are associated with marriage.

Here are some of those that spouses have or are entitled to:

- Visitation rights and the ability to make medical decisions, unless otherwise specified in a living will.
- Benefits for federal employees—many of which are also offered by private employers—such as sick leave, bereavement leave, days off for the birth of a child, pension and retirement benefits, and family health insurance plans.
- Some property and inheritance rights, even in the absence of a will.
- The ability to create life insurance trusts.
- Tax benefits, such as being able to give tax-free gifts to a spouse and to file joint tax returns.
- The ability to receive Medicare, Social Security, disability, and veteran's benefits for a spouse.

- Discount or family rates for auto, health, and homeowners insurance.
- Immigration and residency benefits, making it easier to bring a spouse to the United States from abroad.
- Visiting rights in jail.

Social scientists have also found many positive benefits for married couples and families, including fewer incidents of poverty and mental health problems in families where the parents are married rather than simply cohabitating. Many studies also support the idea that children living with married parents do better in a variety of ways than children in any other living arrangement.

WATCH FOR BROKEN GLASS!

At the end of the ceremony in a Jewish wedding, the groom breaks a glass (for safety considerations, sometimes a lightbulb is used). This gesture symbolizes the destruction of the ancient temple in Jerusalem and represents the couple's identification with Jewish history and tradition. Some also joke that this is the last time that the groom will be able to put his foot down.

From 1970 to 2010, the number of American women heading to the altar tumbled from 76.5 million to 34.9 million. That dip in marriage occurrence isn't indicative of cases of cold feet sweeping the nation, but rather a shift in how people approach the institution. Specifically, unwed couples are choosing more often to live together first and tie the knot later.

Dismal divorce statistics during those intervening decades may have also subliminally instilled concerns about the veracity of 'til-death-do-us-part. After all, romantic love isn't expected to burn brightly forever, at most smoldering into what psychologist Theodor Reik termed a "warm afterglow" in his 1944 book *A Psychologist Looks at Love*. Today, anthropologist Helen Fisher gives young lovers a brief four or five years before the honeymoon phase crests and wanes, correlating to the amount of time it takes to wean a child.

Is that brief window worth the $28,000 spent on the average wedding? Of course, there isn't a correct blanket answer for everyone. Fortunately, contemporary researchers peeking into the interpersonal and neurological pathways and logjams of long-term love have uncovered a host of ingredients for making it work. The only trouble is that some are completely out of our control, at times rendering us defenseless to the whims and wherefores of romance.

⇒ Other Types of Marriage

POLYGAMY

Polygamy, the "state or practice of having more than one spouse simultaneously," was outlawed by the federal government in 1862, as noted in *Black's Law Dictionary*. Still, polygamy exists in the United States, though usually secretly and with great stigma attached. One can also find polygamy in popular culture, such as in the television show *Big Love*, which aired for five years.

In the United States, polygamy is frequently associated with the Church of Jesus Christ of Latter-Day Saints, or Mormonism, although the Mormon church condemns the practice, which is also called plural marriage. An estimated 30,000 to 60,000 polygamists live in Utah and the surrounding states, most of them fundamentalist Mormons. We'll explore more about how polygamy works in a later section.

CIVIL UNIONS, DOMESTIC PARTNERSHIPS, GAY MARRIAGE

A civil union is a legal entity that offers same-sex couples the same legal rights and benefits as marriage. Civil union laws have been proposed in various state legislatures, and now thirty-two states have legalized gay marriage, along with the District of Columbia and ten Native American tribes.

California allows domestic partnerships for same-sex couples and for opposite-sex couples in which one partner is at least sixty-two years old, and many employers and universities offer domestic partnership benefits as well. Like civil unions, domestic partnerships offer many of the same legal benefits that opposite-sex married couples receive, including inheritance and hospital visitation rights. Currently there are over ten U.S. states that allow domestic partnerships.

Before Hawaii allowed civil unions for same-sex couples, it had "reciprocal beneficiary relationships," which are registered through the state's Department of Health. Reciprocal beneficiary relationships are only available to people who are at least eighteen years old and prohibited from marrying by state law, which includes same-sex couples but can also mean a brother and sister or aunt and nephew. Two people who enter a reciprocal beneficiary relationship aren't automatically considered a couple. They simply gain many of the legal rights afforded to married couples, which, for example, could be important to a brother and sister who are supporting each other financially.

COMMON LAW MARRIAGE

For a couple to be considered to have a common law marriage, they must live together, agree that they are married, and present themselves as a married couple, such as by changing their last names and filing joint tax returns. If you've heard that any couple that resides together for seven consecutive years is automatically considered common law husband and wife, that's not true. The amount of time that a couple has to live together is not defined in any state, but if a couple satisfies many of the conditions for common law marriage and lives in a common law marriage state but does not want to be married, they must somehow make that intention clear.

Finally, though a couple who has a common law marriage doesn't have a marriage license or certificate, they still must go through the legal process of obtaining a divorce in order to end their marriage.

WEDDING CUSTOMS

Numerous fascinating wedding customs are enjoyed in cultures around the world. Many cross-cultural similarities can be seen, as marriage traditions frequently symbolize fertility, good health, good luck, or the beginning of a new life. Here are a few interesting customs:

+ In Greece, brides might carry a lump of sugar on their wedding day, symbolizing a sweet life, or ivy, representing endless love.

+ In Norway, friends of a couple plant palm trees on

both sides of the front door of the couple's home as a symbol of fertility.

+ A South African tradition has the parents of the bride and groom carry fire from their homes to the fireplace in the couple's home, signaling the beginning of their life together.

+ It's considered good luck for Venezuelan newly-weds to sneak away from their wedding reception without saying good-bye.

+ In Fiji, the groom is expected to give the bride's father a whale tooth.

BETROTHED THROUGH THE CENTURIES: A TIMELINE OF MARRIAGE

Today's version of marriage looks very different from the unions enjoyed by our ancestors. For most of human history, marriage was more akin to a business deal between men, and the bride in question had very few rights or other options. So how did marriage become associated with love?

The BC Years

In the centuries prior to the first millennium AD, marriage was a good way to ensure your family's safety. By marrying a daughter off to a fellow from a nearby tribe, you expanded the circle of people who you could rely upon in times of famine or violence. Marriage came to be respected as an institution, so much so that people who didn't marry were penalized outcasts. But marriage wasn't so respected that you couldn't escape its bounds once in awhile. In fact, men were expected to have romantic dalliances with mistresses (or even young boys) while maintaining a marriage for purposes of child-rearing. Women were married very young, while men tended to be a little older, and almost all marriages were arranged.

Sixth Century through Seventeenth Century

No Valentines or romantic weekends shared between spouses were to be found in this chunk of time. Marriages continued to be arranged affairs, particularly useful for solidifying status, wealth, and power. Men of one family would present a potential bride to another family, and then they'd negotiate a dowry, or bride price. When the deal was struck, the men presented the bride-to-be with a ring to celebrate the successful transaction. Of course, giving rings to celebrate betrothal has become much more romantic (and expensive) in recent times. Also, during this time, Christian churches began to take a more active role in the marriage process.

Twelfth Century to Thirteenth Century

The union between a man and a woman is described in the sacred texts of most religions. For many centuries, though, the Christian church took a decidedly hands-off approach to marriage. During the twelfth and the thirteenth centuries, however, the church became more involved in performing ceremonies and dictating who could get married.

Churches prohibited marriage between in-laws, blood relations, and families who were linked by the bond of godparent and godchild. The church would often undertake investigations to assure that these conditions were met. It wasn't until the twelfth century that a priest would participate in a marriage ceremony, and it would take another hundred years before the ceremony was actually performed by a priest.

Sixteenth Century

During the Protestant Reformation, men tried to loosen the church's grip on marriage and put the institution in the hands of the government. Protestants also changed the rules about marrying someone with the same blood or marrying someone in your spouse's family. In response, the church dug in its heels and claimed that true marriages required a priest and two witnesses. No matter the religion, however, women were still treated as property possessed by their husbands. Any money in their purse or land held by their family was considered to belong to the man.

Eighteenth Century and Nineteenth Century

Slowly but surely, the idea of being in love with the person you married was beginning to take hold, particularly in England and in France. Though marriage still required women to cede their property and their income, they had a greater chance of giving it to a man they actually cared for. It wasn't until 1870 and 1882 that England passed the Married Women's Property Acts, which allowed women to keep money they earned and inherit property. Marriages were still often arranged, but they began to be preceded by extensive courtships.

1920s to 1950s

With the rise of the automobile, dating became enormously popular. Cars allowed men and women to consider a wider range of marriage prospects, rather than settling for someone nearby. During the 1950s, married women became particularly obsessed with having the perfect home, a phenomenon referred to as "the cult of domesticity." Television programs of the era showed wives and mothers baking pies, vacuuming the home, and putting dinner on the table promptly at 6 p.m.—all while wearing pearls and high heels.

1960s through Today

The rise of second-wave feminism in the 1960s and 1970s brought an end to the cult of domesticity. More women began to work outside the home, and the idea that married women were expected to be subservient to their husbands was considered preposterous. For the first time, women didn't feel that they had to get married, and since then, people have been marrying later, if at all.

In the United States, the government began to loosen some restrictions about who could get married. The 1967 Supreme Court case of *Loving v. Virginia*, for example, eliminated laws prohibiting interracial marriage. However, some restrictions remain, particularly related to same-sex marriage. Today, many countries are grappling over whether to allow same-sex couples to marry. In some states and countries, civil unions between gay couples are permissible, while some states and countries are allowing these couples to legally marry.

MARRIAGE MULTIPLIED: HOW POLYGAMY WORKS

Today, most Americans think of monogamy as the "normal" form of marriage. But as it turns out, strictly monogamous practices are in the minority. In fact, cultures that practice some form of polygamy outnumber monogamous cultures by the hundreds. Some critics suggest that the Western practice of frequent divorce and remarrying represents a form of serial polygamy, though most anthropologists consider it serial monogamy—no one gets married to more than one person at one time.

The Nyinba people of Nepal practice fraternal polyandry. Polyandry is a form of polygamy in which one woman has multiple husbands. In Nyinbian culture, when a woman marries a man, she marries all of his brothers, too. All of the brothers have equal sexual access to the wife, and the entire family cares for the children, although the family may recognize individual brothers as the specific father of a given child. This kind of marriage structure concentrates the wealth and resources of all the brothers into one family, as well as their parents' land and wealth.

Polygyny, on the other hand, rewards males who have access to greater wealth and resources than others. It takes a lot of work and money to support a large number of wives and the children they produce. In biological terms, such a man is an excellent choice for reproducing and passing his genes on to the next generation, which could be expected to be similarly successful. A man can father many children in a short period, while a woman is limited to one pregnancy every nine months. If a successful man has many wives, he can pass on his genes more often.

This is also an advantage in societies where rapid and frequent reproduction is vital for survival. Early Jewish doctrine encouraged polygyny because Jews were a minority and needed to increase their numbers rapidly. Some orthodox Jewish sects advocate polygyny today, and some scholars believe that the Talmud contains passages suggesting tolerance or even encouragement of polygyny.

Islamic tradition addresses polygamy directly. The Koran states that a man is allowed up to four wives, but only if he can support them and treat them all equally. Many Islamic societies continue to allow polygamy, but usually only the most affluent men can afford multiple wives. Westernization has led many younger Muslims to view polygamy as old-fashioned.

In Vietnam, polygamy is not legal, but there's a practical

reason for its practice—decades of war have left the male population severely depleted. Polygamy was also common in China before Confucianism, which supported the practice, fell out of favor. Many African tribes, Native American tribes, and pre-Christian Celts practiced polygamy, often without the conservative restraints on the sexual aspects of it that characterize Mormon polygamy.

MORE ON POLYANDRY AND POLYAMORY, BIGAMY, AND "SWINGERS"

While polyandry is rare, societies that allow both multiple husbands and multiple wives are even rarer. The Amazon Zoe tribe is a notable example of this practice.

Polygamy is often confused with polyamory and bigamy. Bigamy occurs when one man illegally marries more than one woman. He might marry a second wife before his divorce is complete, for example. In rare cases, men have carried on double lives, marrying two women and supporting two families, with neither wife knowing about the other. Practicing polygamy in a country with laws that forbid it is technically bigamy.

Polyamory is the concept of being in love with more than one person at one time. Polygamists can practice polyamory, but polyamorists don't necessarily practice polygamy. They may have living arrangements in which multiple adults form one family, share economic burdens, care for children, and share sexual access with one another. However, they do not usually attempt to form a legal marriage. The line between a polyamorous family and a polygamist family is a fine one. Although broad generalizations always have exceptions, polyamorists tend to have communal, liberal views, while polygamists generally come from conservative religious backgrounds.

Swingers are married people who openly engage in sex with people other than their spouses. They don't generally form lasting relationships beyond friendships, nor do they form family structures. The focus is on sex alone.

Mormons, Polygamy, and the U.S. Legal System

For many people, polygamy is associated with Mormonism, a religion founded by Joseph Smith, Jr., in the early 1800s. Smith claimed to receive messages from an angel, which directed him to golden plates that told the story of ancient people who came to North America from Israel. He supposedly found these plates in New York and translated them into a new scripture known as the *Book of Mormon*.

The *Book of Mormon* itself does not contain specific information about polygamy. However, Smith began to practice polygyny in the 1830s and secretly told his "inner circle" that he had received a revelation: a man should take multiple wives in order to become a king in Heaven. He pointed to Old Testament figures like Solomon and Jacob as examples.

Although his polygyny became well-known and contributed to the persecution of Mormons, Smith never publicly acknowledged the practice, and it wasn't official church doctrine. However, the practice spread and was encouraged by Smith's successor (Brigham Young) in the Church of Jesus Christ of Latter-Day Saints, the "main" branch of Mormonism. By 1852, so many Mormons were practicing polygyny that the church acknowledged it in an official announcement. Mormons refer to polygamy as "plural marriage." By this time, the church had splintered. One branch followed Smith's son, Joseph Smith III, and rejected polygamy entirely.

In 1890, after decades of conflict with the federal government, the leadership of the Church of Jesus Christ of Latter-Day Saints announced that another revelation would change church doctrine. Polygamy became a major obstacle when Utah, where the Mormons settled, applied for statehood.

The new directive forbade polygamy within the church, but it never removed Smith's polygamy revelation from the Mormons' holy texts. Church members who continued to practice plural marriage were excommunicated. This change resulted in another schism, and several small groups of Mormons split off to found their own sects so they could continue to practice polygyny. These groups sometimes refer to themselves as Mormon fundamentalists.

Mormons who practice polygyny today still believe that Joseph Smith, Jr., received a divine revelation that only men who took at least three wives would become "gods" in Heaven. Women who refuse to marry into polygynistic families will be denied entry into Heaven. Modern polygamist Mormon families are characterized by large numbers of children and anywhere from three to more than fifteen wives. Joseph Smith, Jr., is thought to have had as many as forty-eight wives. According to its practitioners, this arrangement has several benefits:

♥ A built-in deterrent to infidelity on the part of the husband. Since he has so many wives to choose from, he won't need to stray from his marriages.

♥ Extra hands to help care for children.

♥ Adherence to a divine directive to lead them into Heaven.

However, they also acknowledge certain difficulties. Jealousy between wives is often a problem, because it can be difficult for the husband to devote enough attention to each wife. Many families create a schedule that regulates which nights the husband sleeps with each wife. Usually, the entire family lives together, with the wives having separate rooms. In some cases, wives might live in separate houses, or the family might live in a duplex.

The economics of polygamy can be hard on the families, as well. Colorado City, Arizona, a strict polygamist enclave, suffers from severe poverty. The families are simply not able to make enough money to support all their wives and children. They rely heavily on welfare and, in some cases, commit welfare fraud. The problem is so severe that Colorado City and similar communities put a serious strain on state welfare systems.

The status of polygamy in the United States is simple: it's illegal. Several federal laws passed in the 1800s made polygamy illegal in all U.S. territories, including Utah. When Utah became a state, legislators included a specific ban on polygamy in the state constitution. Today, few states have laws that specifically outlaw polygamy, but all outlaw bigamy. However, few polygamists try to legally marry more than one wife. They may marry other wives in church ceremonies, but no marriage license exists. Some polygamists marry and then divorce all but one wife, but continue living and sleeping with all of them.

Very few polygamists have been prosecuted. When they are, the prosecution usually focuses on an ancillary offense, such as child abuse or welfare fraud, rather than polygamy itself. A series of raids in the 1950s in which police arrested polygynist husbands resulted in a public relations disaster as people reacted to images of wives and children left without their fathers.

Most polygamists are careful not to "officially" marry more than one wife at a time, so a bigamy conviction is virtually impossible. Laws against cohabitation are vague, hard to enforce, and probably unconstitutional in any case. Any direct prosecution on grounds of polygamy would require one of the wives to act as a witness against the husband. Even if the evidence were easy to come by, there are so many practicing polygamists in Utah and nearby states that the state doesn't have enough money to investigate, try, and jail them all.

THE DARK SIDE OF POLYGAMY

The terms "Mormon" and "Mormonism" are sometimes controversial, as not all sects use this term to describe themselves, and not all sects practice polygamy. In common usage, anyone who follows the teachings of Joseph Smith, Jr., is a Mormon.

It's almost impossible to find hard statistics about polygamy, because plural marriages are rarely documented. Many Mormon fundamentalist sects are closed communities that shun contact with non-members. So it's difficult to determine the frequency of abuses such as marrying minors, marriages between close relatives, or physical and sexual abuse. However, a wealth of anecdotal evidence suggests that these abuses often occur.

In addition to such traumatic abuses, people who have left (or, in their words, "escaped") polygamist families point out that the structure of such families crushes female independence. Husbands have absolute authority, and wives and children are completely subservient to them. Because wives are so dependent on their husband and the other wives, they often lack the life skills to live on their own. This makes it especially difficult to leave. In addition, many polygamist wives were born and raised in polygamist families. They have been in the polygamist lifestyle from birth, so they have a hard time seeing a way out.

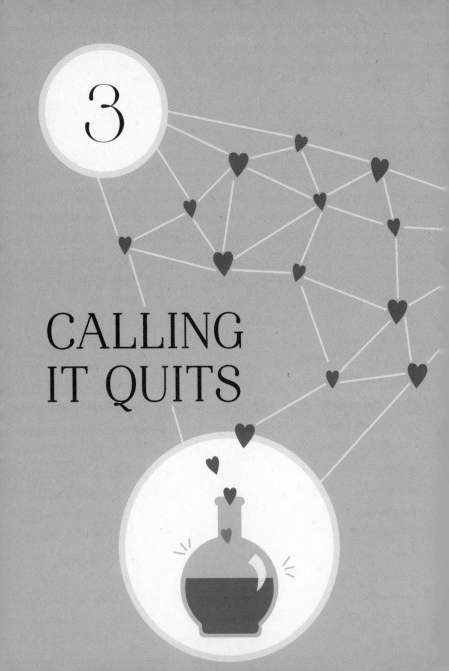

3

CALLING
IT QUITS

ACHY BREAKY HEART: HOW BREAKUPS WORK

E lizabeth Taylor and Richard Burton would've made a compelling case study of the neurological and psychological underpinnings of breakups. The old Hollywood couple initially set the tabloids smoldering when they were caught in flagrante during downtime from filming *Cleopatra* together. Taylor cut ties with her then-husband Eddie Fisher and tied the knot with her Welsh leading man in 1964. Ten years later, Taylor and Burton divorced, only to get remarried a year later and divorced again for good in 1976.

One wonders how their psyches could've weathered the romantic ups and downs in such quick succession. Especially since many adults rate relationship breakups among the worst events of their lives, the Burton-Taylor double divorce seems like the emotional equivalent of a hurricane on the heels of a tornado. At the same time, though, the pair remained close friends even after the final divorce, and Taylor remarried two more times, which also implies that they were somehow better equipped than some other people for saying good-bye to past love. After all, everyone handles breakups a little bit differently.

So how can you tell whether you're equipped to handle a romantic roller coaster? To get started, let's take a bird's-eye view of breakups and find out how, when, and why they tend to happen.

⋛ Anatomy of a Breakup

Is there ever a good time to break up with someone? Is one day of the week more amenable to calling it quits than others? According to data compiled from Facebook status updates in 2009, the most common day to pull the plug is the first Monday in December. That statistical torrent of turmoil is likely explained by people's resistance to breaking up during the holidays. They prefer doing the dirty work beforehand over having to endure a season's-worth of pecking under the mistletoe. Aside from breakup announcement spikes in early December and early March, when many college students take a week off for spring break, the rate of relationship dissolution remains fairly steady throughout the year.

As the world has gone mobile, so have breakups. Consider this generational difference in how the bad news is delivered: men and women born before 1975 will break up with a significant other in person 74 percent of the time, while younger heartbreakers born after 1984 only do so

47 percent of the time. Generation Y is more likely to call someone up (30 percent), send a searing instant message (14 percent), or type out an e-mail (4 percent). Because of this technological interference bereft of interpersonal sensitivity, some psychologists warn that romantic rejection stings more acutely for young lovers.

The rationales for breaking up aren't as easily boiled down to sterile statistics, however. Not surprisingly, cheating is one of the most common relationship deal breakers, along with—and possibly related to—sexual dissatisfaction. One study from Lewis & Clark College in Portland, Oregon, analyzed detailed accounts of breakups and found a macro-level divergence in how men and women decide to break up.

Whereas female respondents tended to present itemized lists of grievances, such as wanting more time together, loyalty, and support, men's explanations were more nebulous. What they missed was an inexplicable, magical quality of bonding and romance. On the micro level, a snapshot of breakup-related Twitter updates in 2009 cited the economy, politics, jealousy, boredom, and even vocal pitch as the final straws for various couples.

Once that Band-Aid rips off whenever and for whatever reason, how does the psyche handle the heartache?

≥ The Emotional Mechanics of Heartache

The psychological symptoms of a breakup aren't pretty. Relationship psychologists identify a spectrum of negative effects, including anxiety, depression, loneliness, and suicide. Moreover, those on the receiving end of a breakup understandably experience a steeper mental free fall. Recovering from the blow isn't a quick process, either. Eight weeks after getting dumped, 40 percent of people in one study exhibited signs of clinical depression, and 12 percent appeared moderately or severely depressed.

Humans' profound emotional reactions to splitting up don't reflect an evolutionary weakness. Rather, it's a visceral response rooted in our mammalian drive toward social bonding that helps guarantee species survival and reproduction. That primitive compulsion undergirds the formation of romantic attachments, and how a person attaches to a partner partially determines how well he or she will manage a breakup from a psychological standpoint.

Just as men and women exist somewhere along a sexual spectrum that encompasses both opposite- and same-sex attractions, they're also scattered along a range of attachment styles. At one end sits anxious attachment, characterized by relational neediness and insecurity, and at the opposite is avoidant attachment that dodges commitment and openness. Anxiously attached partners have the most difficulty accepting breakups and are more likely to turn to unhealthy coping mechanisms, such as drugs and alcohol, to soothe their distress. Avoidant types may simply cut ties with little care for providing closure.

Despite stereotypes of women as the clingier partners in heterosexual couples, attachment style, rather than gender, is more predictive of how strongly someone responds to breakups. Also, considering that women initiate two-thirds of divorces, in a way, that gives them a statistical edge over men in terms of getting over relationships. And along with the psychological advantage of calling the breakup shots, women also tend to have more finely tuned emotional intelligence that may alert them to relationship red flags before men pick up on the signs of danger ahead.

But while people might be able to prepare the heart for hard times, the brain has a mind of its own.

BREAKUP BEST PRACTICES

Breaking up with someone can be uncomfortable for both parties, but doing it in a sensitive, thoughtful way can mitigate the emotional backlash. Elizabeth Svoboda at *Psychology Today* recommends doing it face to face, avoiding blame, and acknowledging

positive aspects of the relationship. As quick and easy
as it may seem, sending a mean text message isn't
considered acceptable breakup etiquette.

This Is Your Brain on a Breakup

To the brain, getting dumped is the pain equivalent of
getting burned by a hot cup of coffee. A 2011 study conducted by a team of neurologists at the Einstein College of
Medicine found that merely looking at a photograph of an
ex-partner energized the neurological regions—the second
somatosensory cortex and dorsal posterior insula, to be
precise—that also process physical discomfort. Defensively,
the dejected brain also signals the release of the stress hormone cortisol and amplifies the body's immune defenses as
though warding off emotional pathogens. Indeed, as additional research confirms, matters of the heart and mind are
intimately connected.

Compare fMRI scans of people recovering from recent
breakups and those of people overcoming a cocaine addiction, and the irrational behaviors that go along with breakup
coping become even more understandable. In other words,
getting over a relationship engages the same neural circuitry
as overcoming an addiction, which is why the absence of lost

loves is felt so potently that it stimulates literal cravings for their presence.

In people who had been dumped, looking at photographs of former romantic partners stimulated their brains' reward systems, which initially secreted pleasure-inducing dopamine at the sight of those breakup initiators in anticipation of their company. But the sad recognition that an ex-partner won't be coming around anymore deprives the reward system of its stimulus, or love drug, kick-starting the ventral tegmental area (VTA) and nucleus accumbens in the central brain.

That duo triggers the motivational urge to possibly see the person and also reanalyze the benefits and drawbacks associated with the relationship—as manifested by the rehashing of past events people often engage in while processing a breakup. And in a domino-like effect, that unsatisfied reward system trips the nearby prefrontal cortex, which elicits feelings of frustration and anger.

On a more positive note, the fMRI data also revealed that the sting of heartache eases with time. Follow-up brain scans months after breakups found lowered levels of activity in regions associated with romantic motivation. By the same token, that also underscores the hard reality that there's no quick and easy route to dissolving amorous attachments. And during that challenging recovery period, it's often tempting to satisfy that neurological craving and rekindle the flame.

Let's Get (Back) Together

During the initial throes of post-breakup angst, the quickest route back to happiness might lead straight into the arms of the most recent ex-partner. Re-evaluating life without someone special in it can burnish the positive aspects of a relationship and push the negative patterns into the background. Missing the sexual intimacy that comes with longer-term mating can also cause couples to rethink whether staying apart is really that wise. Certainly, it took Elizabeth Taylor and Richard Burton some practice to part ways.

Although fixating on getting back together with an ex-partner can be unhealthy, slowing the psychological process of relinquishing the romance, it's happening more often these days. According to a 2010 study at Texas State University, San Marcos, about 20 percent of adults will engage in an on-again, off-again relationship with a significant other over their lifetime.

Young folks are even more likely to change their minds about leaving someone behind, with as many as 60 percent of teenagers breaking up and getting back together down the road—and doing so twice, in most cases. This represents a sharp uptick from related research in the late 1980s and early

2000s, which estimated a get-back-together frequency of 3 to 40 percent. Based on research on motivations for reviving relationships, that tendency toward mending fences might imply greater insecurity among up-and-coming dating pools.

For instance, a 2011 study by the University of Texas at Austin assessing people's reasons for getting back together with an ex highlighted a common theme of relational ambiguity. Along with the expected lingering sentiments, participants noted misunderstandings about the ramifications of a breakup and its negative impact on the couple's post-split relationship as grounds for giving it another shot. But on-again, off-again relationships may inherently restart at a disadvantage, according to related research.

On average, men and women involved in cyclical romances report more negative aspects than positive ones, particularly communication problems and instability. That isn't to conclude that cyclical relationships are doomed for failure, but to emphasize that getting back together doesn't dissolve past problems. Instead of framing it as a fresh start, it would be more accurate to consider reuniting as a redefinition of the existing relationship, warts and all.

Whether or not someone decides to grant an ex a second chance, moving on from the emotional trauma inflicted by a breakup is crucial. The worst thing you can do is to string someone along for your own emotional benefit.

LET'S BE FRIENDS

Same-sex couples are better at staying friends after romantic fallout, compared to opposite-sex couples. According to a 2002 study, lesbian couples are the most likely to maintain post-breakup friendships, followed by gay men. Straight people, it seems, prefer to cut and run.

Moving On from a Breakup

Trite but true, the only cure for breakups is time. It's impossible to wrestle a romance-addled brain into submission and silence the echoes of past love at will. But as mentioned earlier, the brain's reward system gradually stops craving the presence of an ex-partner, and life returns to normal for all intents and purposes. In that way, adjusting to post-breakup existence centers on regaining the sense of self that was absorbed as a natural byproduct of coupling.

A 2010 study from Northwestern University found that the longer the relationship, the more change participants anticipated experiencing post-breakup. Put another way, moving on from a breakup involves unraveling the "we" identity and finding the "me," a process academics are investigating further.

That personal evolution instigated by a significant breakup isn't necessarily a bad thing. In fact, heartbreak and recovery can lead to positive developments that may leave people better prepared for future romance. In seeking out the health-promoting aspects of a breakup, newly singles might pursue hobbies or fitness goals, or reconnect with friends. Psychologists refer to that type of pain-fueled progress as "stress-related growth."

However, a 2003 study involving college students who had experienced difficult breakups found that post-breakup self-cultivation doesn't apply equally to everyone. For instance, rejected women reported the greatest breakup-fostered development. Perhaps by having to overcome the steeper emotional hurdles associated with getting dumped, as opposed to initiating a breakup, those women were more determined to positively learn from the heartbreaking incident.

On the bright side, the reality of breaking up ultimately isn't as bad as we might imagine. In 2007, psychologists at Northwestern University asked participants who were in romantic relationships at the time to forecast the emotional fallout if a breakup occurred, then followed up with them soon after the affairs ended. The men and women turned out to be on the mend much sooner than predicted.

Counterintuitively, the more in love someone is, the

better that person fares post-breakup, compared to his or her expectations. In that case, even romantic diehards can take heart that while falling out of love often comes with a hard landing, humans' emotional immune systems will quickly kick in to heal the cuts and bruises and get us on our way.

REBOUNDING MIGHT BE GOOD FOR YOU

Despite the negative connotations of the term, rebounding after a breakup might not be such a bad idea. For people who exhibit anxious attachment patterns in particular, rebounding can facilitate emotional recovery by demonstrating that other potential romantic partners are out there.

CAN YOU DIE OF A
BROKEN HEART?

W hen a long-time spouse dies, it isn't uncommon for potentially life-threatening health problems to arise in his or her partner soon afterward, or for chronic conditions to take a grave turn. Could broken hearts be to blame?

What happened to an older woman named Dorothy Lee wasn't all that unusual. In 2010, *The Wall Street Journal* reported that after learning her husband of forty years had died suddenly in a car accident, Lee started getting chest pains reminiscent of an impending heart attack. It was as though her body was revolting against the unexpected loss.

Potentially life-threatening health problems like Lee's aren't uncommon in a surviving partner. In fact, anecdotal evidence cites husbands and wives who inexplicably die within weeks, or even days, of each other, and empirical studies have supported that phenomenon. Separate studies involving thousands of couples in Scotland and Israel concluded that the risk of death among widows and widowers surges anywhere from 30 to 50 percent during the first six months after their beloveds pass. After that initial period of bereavement, the statistical risk of death diminishes.

This type of extreme mind-body connection appears to be more common when spousal death is unexpected—as in the case of Dorothy Lee's husband—and the surviving part-ner is ill-prepared to forge on alone. A 1996 study of 158,000 Finnish couples found the highest incidence rate of excess mortality—or statistically unforeseen death—correlated to the accidental, sudden passing of one spouse. Medical doc-tors have attributed that pattern to chronic health problems, psychologists to grief-induced stress, and social workers to lack of a support system. Romantics, meanwhile, might sum it up more sweetly as the byproduct of a broken heart. In some cases, they might not be so far from the truth.

At least, that's what Dorothy Lee experienced on the day her husband died. It turned out that her chest pains weren't the result of blocked arteries, but rather a condition called takotsubo cardiomyopathy, or broken heart syndrome.

Love Hurts, Literally

Japanese doctors first identified broken heart syndrome among five patients in the early 1990s. That cluster stood out among the 415 other heart-attack victims they exam-ined, since the five patients appeared to have no blocked arteries, the usual source of such incidences, and they recovered more quickly and easily than the rest of the study population.

Upon closer inspection of these five patients, the Japanese physicians noticed the hearts' left ventricles had ballooned, resembling a takotsubo, or pot used for catching octopi. That swelling exerted additional pressure on the heart, which explained the temporary heart attack-like symptoms. In 2005, two additional studies on takotsubo cardiomyopathy linked these types of faux heart attacks to extreme emotional states of grief, anxiety, and stress, which helped earn the condition its Shakespearean nickname.

Broken heart syndrome is just one rare example of how heartache can affect our health. Even in minor doses, the sting of rejection and loss doesn't reside solely in the mind but literally travels throughout the body. The brain processes it as a form of physical discomfort and also triggers the inflammatory stress responses that might cause our palms to sweat, hearts to race, and breathing to quicken.

Brain imaging technology, for instance, has revealed that the neurological pathways stimulated by the heartache of being dumped are identical to the pain of a bare hand holding a piping hot cup of coffee. As a result, breakups have been known to incite a host of ailments, including asthma attacks, insomnia, flu, and shingles.

Although parting ways with a romantic partner isn't

necessarily a precursor for broken heart syndrome, extreme stress can aggravate the heart. In response to stress, the body releases catecholamines, or hormones produced in the adrenal glands that prompt our self-protective fight-or-flight reaction. Doctors suspect that a flood of epinephrine and norepinephrine (a.k.a. adrenaline and noradrenaline) briefly disable the heart's muscular cells and slow its pumping functions down to crawl, similar to a heart attack.

Once the condition is correctly diagnosed, patients generally recover—sometimes with the assistance of aspirin and other mild medications to regulate blood circulation—in roughly a week. What medical researchers haven't figured out yet, however, is why women's hearts tend to "break" more often than men's.

Are Women's Hearts More Susceptible to Breaking?

Broken heart syndrome is a relatively rare condition, affecting between 1 and 2 percent of all patients who undergo cardiac catheterization, a diagnostic test for heart problems. In 2007, 89 percent of the 6,230 nationwide reported cases of takotsubo cardiomyopathy were attributed to female patients. According to 2011 research presented to the American Heart Association, women over fifty-five are 2.9 times more likely to develop it than younger females. What remains unanswered is whether women simply respond more severely to extreme stress or whether postmenopausal estrogen levels are determining factors in the syndrome's preponderance among female patients.

The theory that women don't withstand trauma or stress as well as men doesn't hold much water when the range of broken-heart-syndrome triggers are analyzed. It turns out that the condition, probably thanks to its catchy nickname, has been overblown in media reports.

Certainly, there have been cases of left ventricles rebelling in response to death and funerals, but it's more often related to physical dysfunctions, including stroke, acute respiratory failure, burn injuries, and poisoning. Data from the Johns Hopkins University found correlations between emotional loss and broken heart syndrome in only 40 percent of affected patients.

Studies of long-term spouses dying in quick succession also contradict the notion that females are biologically less fit to cope with loss. Statistically, widowers are more likely to succumb following their wives' deaths, whereas widows tend to press on longer.

For all of its doom and gloom, there is a silver lining to broken heart syndrome: it's seldom fatal. Although heart complications occur in approximately 19 percent of cases, the mortality rate for broken heart syndrome rests between 1 and 3 percent. As noted earlier, an expanded left ventricle

will typically return to its normal size, as in the case of
Dorothy Lee, and patients often reach full recovery in as
little as a week—not unlike the metaphorical broken heart
that in time, though scarred, will heal.

BROKEN HEART SEASON?

Whereas heart attacks occur most often during the winter, broken heart syndrome apparently prefers warmer weather. Cases of the cardiac condition spike during spring and summer.

WHY DO BREAKUP SONGS HURT SO GOOD?

I n the face of heartbreak, music can offer incomparable solace. But is listening to breakup songs just
a self-indulgent exercise in sappiness? Or does the
activity actually possess pain-mitigating properties?

For years, Rutgers University anthropologist Helen
Fisher has tirelessly analyzed the evolutionary purposes and
neurological processes of love and its bizarre host of attendant behaviors. In doing so, she and her colleagues have
uncovered how the human brain becomes intoxicated with
the sensation of falling head over heels, and also what happens when that love drug is ripped away. As anyone who's
been through a painful breakup may likely guess, the results
aren't pretty.

Functional MRI brain scans of ten women and five
men who recently had been on the receiving end of a
breakup revealed a flurry of neurological activity at the
mere sight of their ex's photograph. Areas of the brain
associated with addiction, motivation, and anticipation
of loss glittered on the fMRI screens, explaining people's

post-breakup tendencies to feel simultaneously hope-
less, yet longing to see their lost loves and make amends.
Parting ways with a lover is so potently painful that the
brain processes the occurrence the same as it would with-
drawal from a cocaine addiction.

To help participants wipe their mental slates clean of
their angst-inducing exes, the researchers instructed them
to count backward by sevens from 9,247. After forty seconds or so, participants' brain activity returned to neu-
tral, distress momentarily soothed by the mental task of
arithmetic. What Fisher and her colleagues may not have
realized, however, was that there's an easier way to ease
the sting of romantic loss. To quell the breakup blues,
the quickest and most potent prescription often comes in
musical form.

In the face of emotional loss, breakup songs can offer
incomparable solace. Whether someone is in need of a pep
talk on survival from Gloria Gaynor, or a bitter tell-off
courtesy of Bob Dylan, or unadulterated anguish from Sam
Cooke, breakup songs say it all for us, providing space to
wallow and, eventually, motivation to move on. A growing body of research also indicates that listening to breakup
songs isn't just a self-indulgent exercise in sappiness. Music
doesn't just go in one ear and out the other—it actually pos-
sesses pain-mitigating properties.

⋛ Music: A Cure for What Ails

Medical researchers have started paying closer attention to whether more music should be piped into the healthcare process, beyond the Muzak in doctors' offices. Although symphonies and sonatas can't erase pain with the precision of an analgesic prescription, they aren't completely inert, either.

A study published in August 2011 found that breast cancer patients undergoing mastectomies experienced lowered blood pressure and reported lower anxiety when music was playing in the background before, during, and after the operation. Similarly, children treated to post-operative music required less morphine during their recoveries, compared to peers in quieter environments.

Melodious notes serve as an auditory salve because they stimulate the brain's limbic system that regulates pleasurable sensations, such as a rich bite of a cupcake. That network of neurons releases pleasure-prompting dopamine in response to a person hearing his or her favorite jams. Post-breakup, the limbic system is also left hurting. For that reason, putting those heartache songs on repeat can somewhat satiate the brain pathways formerly stimulated by the ex-romantic partner. With the limbic system revived, feelings of hopelessness and depression can ebb as well.

The anxiety of extreme emotional turmoil that comes

with splitting up may also contribute to the satisfaction derived from listening to sad, heart-heavy tunes. For instance, a study conducted at the University of Utah's Pain Research Center found that music-mediated pain relief was most effective among anxious patients. On average, songs don't provide a huge amount of physical pain relief. Study participants reported a mean 0.5-point drop in pain on a 0-to-10 scale when they listened to music. But for people grappling with the climactic anxiety of an impending surgical operation or slogging through romantic despair, music's minor analgesic effect is amplified. In other words, every little note helps.

No wonder the musical library of breakup songs is infinitely extensive, cross-pollinating every possible genre. Since doctors have yet to concoct a cure for heartache, listening to and belting out the blues may just be the next best therapy.

CAN COUNTRY MUSIC DIMINISH DATING PROSPECTS?

Musical tastes can potentially be a relationship deal breaker, according to a 1989 study published in the journal *Communication Research*. A penchant for country music is particularly polarizing, researchers found, souring sexual attractiveness ratings among men and women alike.

TOP FIVE LOVE LIES

P op principles about love often end up on the chopping block, but classic wisdom about happily-ever-after may not be so spot-on, either. Which time-honored, love-related truisms break down when put under the microscope?

1. First Love Is the Best Love

Like riding a bike, witnessing the breathtaking vistas of the Grand Canyon, or finishing a marathon, the euphoria of falling in love is most salient the first time around, many people find. Often romanticized as an emotional high that can never be achieved again, first loves certainly make a deep impression on people's future relationships. But Jennifer Beer, a former graduate student at University of California at Berkeley, found that those initial, heady relationships can do more harm than good.

Since first serious relationships are so self-defining, ones that sour might disable some from pursuing healthy, happy partnerships in the future. In fact, Beer projected that formative romances can have a deeper impact on relationship patterns moving forward than people's interactions with and observations of their parents' partnerships.

Because of the major indentation that first love leaves on the heart, British sociologist Malcolm Brynin advocates approaching initial relationships more casually. His research concluded that those relationships establish unrealistic standards for coupling moving forward, setting people up for dysfunction down the road. Instead of clinging to the notion that the first is the best, pragmatic men and women would be wiser to assume better relationships are in store.

Moving on from that first love and acquiring relationship experience along the way, men and women may return to old flames with surprising success rates. A study conducted in the late 1990s by California State University psychologist Nancy Kalish calculated a 76 percent chance that couples reuniting after at least a five-year hiatus would stick together.

2. Women Are More Romantic

Stereotypically, women are considered the more moony-eyed, love-obsessed sex. Female readers also comprise the majority of the romance novel market, devouring fateful tales of plain-faced governesses, vampire girlfriends, and otherwise whimsical leading ladies. When it comes to real-world relationships, however, emotive women might be more practical than they're given credit for.

A 2011 study conducted at Pennsylvania State University found that male college students were more likely to fall in love and drop the magic three words sooner than their female counterparts. Moreover, a separate analysis of six studies on romantic communication found that men react more positively to hearing "I love you" than women do.

Evolutionary psychology explains that unexpected gender difference as a female tactic to root out a suitable suitor. It's advantageous for them to refrain from falling head over heels and allow men to take the love leap first to mitigate the reproductive risk of being left in the lurch as a single mother. In today's dating landscape, it's also a way to avoid being dismissively labeled as a fool for love.

3. Opposites Attract

To paraphrase Columbia University psychologists who tracked mingling behavior at a cocktail party in 2007: guests at mixers do not mix. As much as people might like to think that they want to meet new and interesting strangers, the knee-jerk compulsion is to flock toward kindred spirits. That Columbia cocktail party study is just one of many confirming a similarity-attraction effect (SAE) among humans that can be readily witnessed in high school cafeterias segregated by teenage cliques.

Yet in romantic scenarios, people start out with earnest intentions to disprove the SAE and support the cliché that opposites attract. For instance, a 2009 survey polled participants about the types of desired qualities in potential mates, and 85.7 percent described personalities different from their own. But they went on to pursue like-minded mates, because SAE also applies to romantic coupling under the anthropological auspices of positive assortative mating.

Cross-cultural research has confirmed that men and women tend to be drawn toward compatible partners from similar socioeconomic backgrounds, education levels, and even political affiliations. That isn't to say that folks are seeking out a veritable twin to spend the rest of their lives with, but rather complementary companions.

4. Monogamy Is Meant to Be

In late June 2011, the *New York Times Magazine* featured a comment-stoking story about infidelity. Rather than railing against statistics of sexual activity outside marriage and long-term relationships, the article provided an avalanche of anecdotal evidence that for today's modern lovers, building a life together doesn't necessarily mean sharing a bed together every night. As readers knit their brows feverishly, wondering whether such an open arrangement could be emotionally sustainable, 95 percent of mammalian species shrugged their shoulders and carried on with their philandering ways.

Though combined social and sexual monogamy is the standard relationship construct among many global cultures, it isn't as nature intended. Only 3 to 5 percent of mammalian species practice lifelong social monogamy, and DNA technology has demonstrated that those few faithful breeds, such as wolves, really aren't so sexually devoted. Researchers also have posited that social monogamy is merely a byproduct of limited options. When nature provides enough desirable mates to go around—hello, *Homo sapiens*—evolution opts for promiscuity.

Accepting the innate human impulse toward sexual exploration—along with the possibility of catering to it within an emotionally monogamous relationship—isn't a rejection of monogamous relationships. Rather, some argue that it's an honest acknowledgment that the incredible diversity of the human species may naturally result in a diversity of pair-bonding styles. In that light, it doesn't sound as much like moral decay as like societal evolution.

5. Breakups Are Unbearable

Presented with the prospect of romantic breakups, people tend to batten down their emotional hatches and prepare for the worst, as if an incoming hurricane were about to render the psyche into an unrecognizable scrap of wreckage. Or, in academic parlance, people make for terrible emotional meteorologists, poorly skilled at affective forecasting. When asked to predict how potently misfortune—lost football games, sour bets, and undesirable election results—will sting, people see nothing but stormy weather ahead, studies have shown.

A 2007 study conducted at Northwestern University also discovered that the possibility of a breakup feels a lot worse than the reality of one. The researchers checked in with twenty-eight participants before, during, and after splits from significant others to gauge how their expectations for dealing with and moving on from the breakups compared with their real-world progress.

Just like people in a separate study, who imagined being let down for much longer about losing a bet than they actually were, the heartbroken recovered more quickly than expected. This might be because brains begin reevaluating whether severed romantic relationships were so sweet after all, neurologically ushering the newly single toward emotional repair faster than they might consciously notice. Sure, the only cure for soured love is time, but it's at least a small comfort that blessed relief will arrive sooner than anticipated.

QUIZ: Calling It Quits: The Achy Breaky Breakup Quiz

Adults commonly rate relationship breakups as some of the most traumatic events of their lives, casting them from the heights of euphoric love into the emotional doldrums. But guess what's one of the best ways to ease the post-relationship blues? Distracting the brain from revisiting old love. How is that possible? By taking this Breakup Quiz, of course. Check your answers in the back of the book on page 199.

1. **Falling in love has a similar effect on the brain as what drug?**
   a. Marijuana
   b. Cocaine
   c. Methamphetamine

2. **Which genre of music is most likely to be a relationship deal breaker?**
   a. Country
   b. Heavy metal
   c. Classical

3. For people born after 1984, which is the second most common means of breaking up with someone?

    a. Telephone

    b. Facebook

    c. Instant message

4. Which of the following is NOT a common month for breakups?

    a. March

    b. July

    c. December

5. Are men or women more likely to break up because of the absence of a "magical quality"?

    a. Men

    b. Women

    c. Neither. People typically have a specific reason for breaking up.

6. What percentage of people exhibit signs of clinical depression following breakups?

    a. 10 percent

    b. 20 percent

    c. 40 percent

7. Which of the following psychological attachment styles tend to weather breakups worst?

   a. Avoidant attachment

   b. Anxious attachment

   c. Secure attachment

8. Which of the following is NOT a physiological reaction to a breakup?

   a. Pain receptor stimulation

   b. Immune system activation

   c. Pupil constriction

9. Which of the following types of couples are most likely to stay friends after breakups?

   a. Opposite-sex couples

   b. Lesbian couples

   c. Gay couples

10. True or False: Couples that get back together tend to be happier than those dating for the first time.

   a. True

   b. False

LIAR, LIAR: DOES CHEATING RUN IN THE FAMILY?

T he public perception of the legendary Kennedy dynasty could be summed up in three words: Catholicism, politics, and of course, affairs. Over the years, the extramarital infidelities of America's royal family have continually piqued tabloid interest into the goings-on behind the Kennedy compound gates. As recently as February 2012, in fact, a former White House intern came forward with her story of an interlude with President John F. Kennedy, whose wandering eye was already widely acknowledged. And the Kennedy canoodling didn't stop at the threshold of the Oval Office.

The former president's married brother Robert F. Kennedy may have engaged in dalliances with screen siren Marilyn Monroe, and some allege that he and Jacqueline Kennedy carried on after her husband's assassination. In 1969, youngest brother Ted Kennedy also raised eyebrows when he abandoned the Chappaquiddick Island drowning site of Mary Jo Kopechne. He'd left a party alone with Kopechne while his wife, Joan Kennedy, was pregnant.

Bobby's daughter Kerry became embroiled in a public divorce from future New York governor Andrew Cuomo when he discovered her long-term affair with a family friend in 2003.

Was there something in the water at Camelot? Clearly, that handful of Kennedys isn't the only cluster of Americans guilty of sexually straying from husbands and wives. Collective data on cheating behavior among married couples projects that between 20 and 40 percent of men and 20 and 35 percent of women have committed adultery against their respective spouses, which is actually a testament to the institution's monogamous bedrock, since 70 percent of dating couples report cheating.

Moreover, certain environmental factors may have predisposed Kennedy family members to extramarital activity. For instance, relationship psychology research has linked flusher bank accounts with higher likelihoods of cheating, since money literally affords more opportunity and access to sex.

But recent scientific findings also point to a more basic explanation for the Kennedys' common habit of cavorting away from the conjugal bed. Perhaps cheating is simply inherited, passed along the branches of the sprawling family tree through their blue-blooded Hyannis Port genes.

⇗ The Infidelity Gene Debate

In October 2011, Czech anthropologists published a study possibly explaining the Kennedy brothers'—Robert's, John's, and Ted's—common tendency to seek sex outside of marriage. They surveyed a group of eighty-six cohabitating couples about their relationship satisfaction and infidelities, in addition to their knowledge of any parental unfaithfulness.

The only correlation they unearthed was between fathers' and sons' cheating behavior. Specifically, a father's known history of extramarital affairs predicted a higher probability that his son would follow in his philandering footsteps. Lo and behold, the Kennedy pater familias, Joseph, carried on his fair share of extramarital romances, including one with cinema icon Gloria Swanson in the late 1920s.

Scientists worth their salt know that such correlative relationships don't prove population-wide causations, but that 2011 Czech study nevertheless supported earlier evidence of a genetic component to infidelity. A 2008 headline-sparking genetic investigation out of Sweden identified allele 334, a genetic variation associated with male infidelity because it interferes with the brain's processing of vasopressin, a neurochemical associated with monogamous pair bonding. Men with two copies of allele 334 were twice as likely to have weathered a major relationship crisis, compared to men bearing one allele copy.

Their romantic partners also reported the lowest relationship satisfaction among the participant pool.

Two years later, a separate team of scientists implicated the brain's dopamine D4 receptor gene, linked to addictive behavior, as another key in the infidelity ignition. One type of D4 genetic variation called a 7R+ allele effectively diminishes the concentration of dopamine receptors in the brain's reward system. Previous studies had likewise associated it with sensation-seeking behavior, including monetary spending, promiscuity, and cigarette smoking.

And, indeed, infidelity study participants possessing the 7R+ allele reported higher numbers of sexual partners than those with a standard 7R− allele. Compellingly, the 7R+ folks were no more prone to cheating than their 7R− counterparts. Once they crossed the adultery line, however, they tended to do so with a higher number of sexual partners.

But those savvy scientists would also admonish against treating genetic predisposition as crystal-ball predictions, despite splashy news stories suggesting otherwise. A genetic study published in November 2010 from Saint Thomas' Hospital in London supports this. The researchers' analysis of 1,600 adult female twin pairs attributed 38 percent of unfaithful behavioral to inheritable genes, discrediting the notion that predisposed cheaters are completely at the mercy of vasopressin- and dopamine-mangling DNA. In that case, the ultimate fate of romantic fidelity—to cheat or not to cheat—is left up to environmental chance and personal choice.

CONCLUSION

⇗ What's the Future of Love: Virtual Sex?

So what is the future of love and sex: virtual love and sex?
Potentially yes. The future is bright for virtual sexuality. But
how can you feel kisses that aren't truly there? How can you
caress something that exists as little more than a dream? How
can we connect carnally with an immaterial world?

Computer scientists today are testing how to transmit
sense data across the Internet and enable interactions with
virtual entities and environments. They're steering us toward
a reality previously only dreamt of: sexual congress between
the material and immaterial worlds.

⇗ Virtual Sexuality Takes Shape

Think of any great artistic or technological achievement, and
chances are it winds up in our pants at some point. Poetry
gave us pornography. The telephone gave us sex hotlines,
and vulcanization gave us the modern condom. It's just part
of who we are.

Today, more than 2.1 billion people surf the World

Wide Web. Internet pornography is a multibillion-dollar industry, and although an exact count of websites is difficult to formulate, we know that at least one hardworking web-filtering service (CYBERsitter) currently blocks more than 2.5 million porn sites.

Roughly speaking, all of our sexual uses of technology can be divided into two categories:

- Sexual communication with another person (sexting, sex chat rooms, and webcam cybersex).
- Sexual interaction with a simulation (interactive sex video games).

All of it is essentially computer-mediated communication, so while you won't find companies such as Microsoft pondering the best way to touch a woman's breast in a video game, the company's Xbox 360 Kinect motion controller was barely out a month before independent developers unveiled a demo for an erotic simulation based on the Kinect's open-source software.

Human telecommunication began with the mere transmission of words and ideas via the telegraph. We quickly moved on to sending sounds and sights. Today, the industry continues to work toward the seamless transmission of our entire sensory experience.

⋛ Physical Touch in a Digital World

They don't call it "carnal knowledge" for nothing. Every lover's embrace is essentially sensory data, but to what extent can we truly digitize, transmit, and receive that information? Let's start with the sense of touch.

Sure, a computer mouse or video-game joystick allows you to manipulate items in a computer environment, but the realm of physical touch falls to the field of computer haptics. Haptic technology generally takes the form of a glove that allows the user to not only control but actually feel virtual items.

Our sense of touch is poorly understood compared to our sense of sight, so the continued development of haptic gloves involves such advanced measures as the use of fMRI brain scans. Hijacking such technology for sexual purposes is inevitable, but far more pressing needs drive funding to computer haptics, ranging from telesurgery carried out over vast distances to virtual reality military training and space exploration.

Of course, sexual touches incorporate far more than the use of our hands, and scientists are already hard at work on haptic technology for our other parts. Just consider the Hug Shirt, a Bluetooth-enabled garment from London-based CuteCircuit that uses embedded sensors and actuators to simulate the warmth and touch of an affectionate embrace.

Suddenly, a transoceanic hug becomes as simple as sending a text message.

Or if you fancy something a little more randy, then take heart in the Kiss Transmitter prototype from Kajimoto Laboratory at Tokyo's University of Electro-Communications. Resembling a cross between an electric toothbrush and a joystick, the device is designed to transmit all the tongue-swirling intensity of a French kiss across the digital divide.

Between the Hug Shirt and the Kiss Transmitter, you'd never have to type "XOXO" at the end of an e-mail again. For now, however, neither technology is commercially available.

As for all the other physical touches that encompass human sexuality, look no further than the field of teledildonics, which largely breaks down to the development of vibrating or otherwise automated sex toys controlled either by a remote user or by a program.

As low tech as that sounds (and generally is), the electronic transmission of physical pleasure takes a rather high-tech turn with Dr. Stuart Meloy's Orgasmatron. Created in 2008, this device sends an electrical pulse through nerves in the spinal cord that inform parts of the brain processing pleasurable sensations in the female genitalia. A study published in the journal *Neuromodulation* even reported that four women who had previously lost the ability to experience orgasms regained it with the device.

In the example of the Orgasmatron, we see that the future of virtual sex may rely not only on haptic gloves and vibrating sensors, but also on the complex manipulation of the human nervous system.

Wake Up and Smell the Future

Smell and taste complete the human sensory experience, and rest assured that research continues into the creation of digitally transmitted smells. Before you scoff at the importance of smell in human sexuality, just remember that about one of every fifty genes in the human genome concerns this most ancient of senses.

Without smell in the mix, how can we expect to fully buy into a virtual experience? And as smell plays such a vital role in our sense of taste, the importance of digital scent technology becomes obvious.

Printable odors follow the same logic as your standard laser printer. While the printer creates a wide variety of colors from a limited palette of primary colors, an odor printer creates a variety of smells out of primary scents.

This brings us back to our sense of vision. While visual rendering technology continues to advance at an amazing rate, the addition of smell, taste, and touch into the virtual reality experience actually lightens the burden for visual simulation. This is due to what scientists call cross-modal attention effects, which govern the way the brain prioritizes some sensory inputs over others.

For instance, the kiss you feel on your cheek takes sensory priority over the texture of the sofa cushion underneath you—and it takes priority at a neurological level. In fact, the amount of brainpower allotted to sight drastically decreases once sound, smell, and touch are introduced.

Researchers at the University of York's Audio Laboratory are working to employ cross-modal attention effects to fine-tune the virtual reality experience, orchestrating the full range of human senses in the required intensities. They've even considered employing a mouthpiece to simulate different textures against the tongue and mouth—for virtual food chewing, they claim.

VIRTUAL SEX CRIMES

In 2001, British satirist Chris Morris duped U.K. Parliament member Barbara Follett (wife of novelist Ken Follett) into speaking out against Hidden Online Entrapment Control System (HOECS) Internet games that allowed pedophiles to touch unsuspecting children through a computer screen. Morris concocted the prank to satirize Britain's pedophilia hysteria, but the idea raises legitimate questions. Will we manage to regulate virtual sexuality effectively in the future, or will technological advancement outpace our abilities to police and protect its users?

The technology continues to emerge, but the day is fast approaching when two lovers may slip into their virtual reality cocoons and fall into each other's arms across intercontinental or even interplanetary distances.

Still others may turn their backs on real-world romance altogether. They'll embrace simulated celebrities, impossible anime bodies, and fetishes previously only imagined. Tomorrow's virtual hedonaut will reach out with a haptic glove, stare into programmed eyes, and touch flesh that exists as little more than a dream.

Whatever may happen, there is no doubt that love and sex will continue to be dominant forces in our lives and in the future of humanity. The difference today is that we have more choices than ever for finding and choosing love. We simply have to take care to choose well and choose wisely.

CONTRIBUTORS

Josh Clark

Cristen Conger

Molly Edmonds

Shanna Freeman

Ed Grabianowski

Robert Lamb

Lee Ann Obringer

Jacob Silverman

Tracy V. Wilson

ANSWERS TO THE QUIZ

⤳ The Achy Breaky Breakup Quiz

1. Falling in love has a similar effect on the brain as what drug?

    Answer: B. Love can feel like a drug because it stimulates the same pathways as cocaine, fMRI scans reveal. That's why coming down from it can feel like withdrawal.

2. Which genre of music is most likely to be a relationship deal breaker?

    Answer: A. A 1989 study published in the journal *Communication Research* found that a taste for country music diminished both men's and women's sexual attractiveness scores among a sample of heterosexual participants.

3. For people born after 1984, which is the second most common means of breaking up with someone?

Answer: A. About 30 percent of Gen Y heartbreakers will pick up the phone to deliver the bad news, followed by 14 percent who prefer to send a sad instant message.

4. **Which of the following is NOT a common month for breakups?**

Answer: B. According to data gathered from Facebook status updates, breakups spike in early December and March. July is generally smooth sailing for summertime lovers, however.

5. **Are men or women more likely to break up because of the absence of a "magical quality"?**

Answer: A. According to a 1986 study published by the American Psychological Association, men were more likely to miss the "magic" and call it quits.

6. **What percentage of people exhibit signs of clinical depression following breakups?**

Answer: C. According to one study, 40 percent of people appeared clinically depressed eight weeks after parting ways with a beloved.

7. **Which of the following psychological attachment styles tend to weather breakups worst?**

Answer: B. Anxiously attached people who tend to be needier in relationships often have the most difficulty accepting and processing romantic breakups.

8. **Which of the following is NOT a physiological reaction to a breakup?**

Answer: C. The brain processes breakups similar to physical pain, putting pain receptors and the immune system on high alert.

9. **Which of the following types of couples are most likely to stay friends after breakups?**

Answer: B. According to a 2002 study, lesbian couples are the most likely to maintain postbreakup friendships, followed by gay men.

10. **True or False: Couples that get back together tend to be happier than those dating for the first time.**

Answer: B. According to research among college

couples who have broken up and gotten back together, giving a relationship a second shot is associated with lowered happiness and satisfaction, possibly because there are more interpersonal problems at the outset.

SOURCES

⇃⇂ Introduction: Why Do We Love?

American Physiological Society. "Love Really Is 'All in Your Head,' Though Intense Romantic Love Looks More Like the Brass Ring Than a Bouquet of Roses." May 31, 2005. www .exercisedaily.org/cgi-bin/details.pl?article_id=1068.

Aron, Arthur. "Brain Network Links Cognition, Motivation." EurekAlert! August 19, 2010. www.eurekalert.org/pub _releases/2010-08/wuis-bnl081910.php.

Aron, Arthur. "Why Do We Fall in Love?" Discovery Fit & Health. Accessed August 30, 2010. health.howstuffworks.com /relationships/love/why-do-we-fall-in-love.htm.

The Economist. "I Get a Kick Out of You." February 12, 2004. www .economist.com/node/2424049.

Gusatella, Adam J., et al. "Intransal Arginine Vasopressin Enhances the Encoding of Happy and Angry Faces in Humans." Biological Psychiatry 67 (June 15, 2010): 1220–1222. www .biologicalpsychiatryjournal.com/article/S0006-3223%2810 %2900243-X/abstract.

Johnson, Steven. "Emotions and the Brain: Love." Discover, May 1, 2003. discovermagazine.com/2003/may/featlove.

Pettifor, Eric. "Beyond Dichotomies: Health and Values in Maslow's Holistic Dynamic Theory." Personality and Consciousness. Accessed August 30, 2010. pandc.ca/?cat=abraham_maslow &page=beyond_dichotomies.

Sing365. "'Nature Boy' lyrics." Accessed August 30, 2010. www.sing365

.com/music/lyric.nsf/Nature-Boy-lyrics-Nat-King-Cole/ADF
31EF29958DA5948256AF1000B59DE.

♥ ♥ ♥ ♥ ♥

"Have We Met Before?": How Flirting Works

De La Vina, Mark. "A Guide to Flirting 4 u—Get the Message?" *Daily Telegraph*, August 22, 2007.

Fox, Kate. "The Flirting Report." Social Issues Research Centre. 2004. Accessed January 14, 2008. www.sirc.org/publik/advanced _flirting.shtml.

Gusmaroli, Danielle. "How to Flirt with Success." *Daily Mail*, February 23, 2006.

H2G2. "Floriography—the Language of Flowers." March 10, 2010. www.h2g2.com/approved_entry/A5268035.

Moore, Monica M. "Nonverbal Courtship Patterns in Women." *Ethology and Sociobiology* 6 (1985): 237–247.

Robson, David. "It's Great to Tease at the Office, But What's Flirty and Not Dirty?" *Daily Express*, February 25, 2007.

Rodgers, Joann Ellison. "Flirting Fascination." *Psychology Today*, January 19, 2006. psychologytoday.com/articles/index.php ?term=19990101-000033&page=1.

Smith, Dinitia. "A New 'Kama Sutra' without Victorian Veils." *New York Times*, May 4, 2002. www.nytimes.com/2002/05/04/books /a-new-kama-sutra-without-victorian-.

Spiegel Online International. "Scoring a German: Flirting with Fräuleins, Hunting for Herren." June 5, 2006. www.spiegel.de /international/0,1518,419712,00.html.

Stepp, Laura Sessions. "Modern Flirting." *Washington Post*, October 16, 2003.

Stevenson, Suzanne. "There's More to Her Flirt." *London Evening Standard*, March 17, 2003.

Williams, Daniel. "When Fingers Do the Flirting." *Time*, May 29, 2006.

content.time.com/time/magazine/article/0,9171,1198940,00
.html

Young, John H. *Our Deportment: Or the Manners, Conduct and Dress of the Most Refined Society*. Project Gutenburg, (1881) 2006. www .gutenberg.org/files/17609/17609-h/17609-h.htm.

♥ ♥ ♥ ♥ ♥

Pheromones at First Sight? The Top Five Physical Signs of Attraction

Akademiai, Kiado. "Experimental Evidence That Women Speak a Higher Pitch to Men They Find Attractive." *Journal of Evolutionary Psychology* 9 (March 2011): 57–67.

Fisher, Helen. "Love at First Sight." *O, The Oprah Magazine*, November 2009. Accessed January 20, 2012. www.oprah .com/relationships/Love-at-First-Sight-Helen-Fisher-Love -Column.

Fisher, Helen E. "Lust, Attraction and Attachment in Mammalian Reproduction." *Human Nature* 9 (1998): 23–52. www .helenfisher.com/downloads/articles/10lustattraction.pdf.

Foster, Craig A. "Arousal and Attraction: Evidence for Automatic and Controlled Processes." *Journal of Personality and Social Psychology* 74 (January 1998): 86–101. Accessed January 20, 2012. psycnet.apa .org/index.cfm?fa=buy.optionToBuy&id=1997-38342-007.

Gueguen, Nicolas. "Mimicry and Seduction: An Evaluation in a Courtship Context." *Social Influence* 4 (2009): 249–255. Accessed January 20, 2012. www.tandfonline.com/doi /abs/10.1080/15534510802628173.

Jarrett, Christian. "Mimicry Improves Women's Speed-Dating Success." *Research Digest of the British Psychological Society*. October 13, 2009. Accessed January 20, 2012. bps-research-digest.blogspot .com/2009/10/mimicry-improves-womens-speed-dating.html.

Landau, Elizabeth. "What Your Heart and Brain Are Doing When You're in Love." CNN. February 10, 2010. Accessed January 20, 2012.

articles.cnn.com/2010-02-12/health/love.heart.brain_1_heart
-rate-stress-hormone-romantic-love?_s=PM:HEALTH.

Martin, Gary. "Imitation Is the Sincerest Form of Flattery." The Phrase
Finder. Accessed January 20, 2012. www.phrases.org.uk
/meanings/imitation-is-the-sincerest-form-of-flattery.html.

Matsumura, Kenta, Takehiro Yamakoshi and Peter Rolfe. "Love Styles
and Cardiovascular Responder Types." *International Journal of
Psychological Studies* 3 (December 2011): 21.

McLoughlin, Claire. "Science of Love—Cupid's Chemistry." The
Naked Scientists. February 2006. Accessed January 20, 2012.
www.thenakedscientists.com/HTML/articles/article/claire
mcloughlincolumn1.htm/.

Murphy, Cheryl. "Learning the Look of Love: In Your Eyes, the Light, the
Heat." *Scientific American*, November 1, 2011. Accessed January
20, 2012. blogs.scientificamerican.com/guest-blog/2011/11/01/
learning-the-look-of-love-in-your-eyes-the-light-the-heat/.

Obringer, Lee Ann. "How Love Works." HowStuffWorks.com. Accessed
January 20, 2012. people.howstuffworks.com/love6.htm.

O'Luanaigh, Cian. "Women with High-Pitched Voices Go Nuts over
Men with Deep Voices." *The Guardian*, July 16, 2010. Accessed
January 20, 2012. www.guardian.co.uk/science/blog/2010
/jul/16/women-high-pitched-voices-men.

Parry, Wynne. "Deep-Voiced Men Don't Have 'Macho' Sperm."
LiveScience. January 02, 2012. Accessed January 20, 2012.
www.livescience.com/17697-voice-pitch-men-semen.html.

PhysOrg. "The Sound of Seduction: Lowering Voice May Be Means
of Signaling Attraction." May 20, 2010. Accessed January 20,
2012. www.physorg.com/news193585517.html.

Silverman, Jacob. "Why Do Old Couples Look Alike?" HowStuffWorks
.com. Accessed January 20, 2012. www.howstuffworks.com
/environmental/life/genetic/old-couples.htm.

Spiers, A.S.D. and D.B. Calne. "Action of Dopamine on the Human
Iris." *British Medical Journal* 4 (November 8, 1969): 333–335.

Accessed January 20, 2012. www.ncbi.nlm.nih.gov/pmc/articles/PMC1629580/.

Tombs, Selena and Irwin Silverman. "Pupillometry: A Sexual Selection Approach." *Evolution & Human Behavior* 25 (July 2004): 221–228. Accessed January 20, 2012. www.ehbonline.org/article/S1090-5138%2804%2900026-1/abstract.

Ulin, Don. "Why Do My Hands and Feet Sweat So Much?" Indiana Public Media: A Moment of Science. January 25, 2011. Accessed January 20, 2012. indianapublicmedia.org/amomentofscience/hands-feet-sweat/.

♥ ♥ ♥ ♥ ♥

⇉ Pucker Up: How Kissing Works

ABC Austrailia: Catalyst. "The Science of Kissing." March 14, 2002. www.abc.net.au/catalyst/stories/s498838.htm.

Ainsworth, Claire. "It Started with a Kiss." *New Scientist*, December 23, 2000. www.newscientist.com/channel/sex/love/mg16822703.700.

Barr, Cameron W. "Japan's Teens Pucker Up in Public." *The Christian Science Monitor*, February 17, 1998. www.csmonitor.com/1998/0217/021798.intl.intl.3.html.

BBC News. "Kissing Many 'Risks Meningitis.'" February 10, 2006. news.bbc.co.uk/1/hi/health/4696974.stm.

Blue, Adrianne. *On Kissing: Travels in an Intimate Landscape*. New York: Kodansha America, 1997.

Bowen, Jon. "Kissing Therapy." Salon. February 14, 2000. www.salon.com/2000/02/14/kissing_2/.

Boyles, Salynn. "'Kissing Disease' Increases Cancer Risk." WebMD. October 1, 2003. www.webmd.com/content/article/74/89355.htm.

Coghlan, Andy. "Kissing the Right Way Begins in the Womb." *New Scientist*, February 13, 2003. www.newscientist.com/article.ns?id=dn3386.

Dugatkin, Lee. "Why Don't We Just Kiss and Make Up?" *New Scientist*, May 7, 2005. www.newscientist.com/article/mg18624981.300-why-dont-we-just-kiss-and-make-up.html.

Foer, Joshua. "The Kiss of Life." *New York Times*, February 14, 2006. www.nytimes.com/2006/02/14/opinion/14foer.html?ex=1297573200&en=64bad474e17f3713&ei=5088&.

Gorman, Kathleen. "A Passionate History: Kissing." *Hartford Courant*, May 8, 1995.

Harvey, Karen, ed. *The Kiss in History*. Manchester, UK: Manchester University Press, 2005.

Kim, Gina. "Kissing: Nature's Cure-All, for Most." *The Seattle Times*, February 12, 2006. seattletimes.nwsource.com/html/nationworld/2002800160_kiss12.html.

"Kissing: Fast Facts." *Science World*, February 7, 2003. connection.ebscohost.com/c/articles/8955172/kissing-fast-facts.

Kong, Dolores. "Though Not Quite the Kiss of Death, Smooching Does Have Hazards." *Detroit Free Press*, February 5, 1989.

Mount Holyoke College. "On Kissing: A Q&A with Michael Penn." www.mtholyoke.edu/offices/comm/news/kissing.shtml.

Roan, Sheri. "Teens' Heated Kisses May Have a Price." *Los Angeles Times*, February 13, 2006. articles.latimes.com/2006/feb/13/health/he-kiss13.

Summerfield, Robin. "Pucker Up: Kissing Keeps Us Happy, Healthy, and Connected." *Calgary Herald*, February 13, 2006. www.canada.com/topics/lifestyle/valentinesday/story.html?id=1183ea35-4773-40a5-90c4-6e4be528c7b5.

Texas A&M University. "Kissing a Favorite Valentine's Day Practice, Says Anthropologist." Press Release. February 13, 2006. business.highbeam.com/138798/article-1P3-1179927091/kissing-favorite-valentine-day-practice-says-anthropologist.

Thomas, Keith. "Put Your Sweet Lips…" Times Online. June 11, 2005. www.thetimes.co.uk/tto/arts/article2400845.ece.

Tiefer, Leonore. "The Kiss: A Fiftieth Anniversary Lecture." The Kinsey
    Institute. October 24, 1998. www.indiana.edu/~kinsey/services
    /tiefer-talk.html.

Tortora, Gerald J. and Sandra Reynolds Grabowski. *Principles of Anatomy
    and Physiology*. Hoboken, NJ: John Wiley & Sons, 2000.

Treadwell, Ty. "Hanging Upside Down for a Kiss." *The Christian Science
    Monitor*, February 26, 2003. www.csmonitor.com/2003/0226
    /p12s01-altr.htm.

♥ ♥ ♥ ♥ ♥

⋛ Sex on the Brain: How Lust Works

Baumeister, Roy F., Kathleen R. Catanese, and Kathleen D. Vohs.
    "Is There a Gender Difference in Strength of Sex Drive?
    Theoretical Views, Conceptual Distinctions, and a Review of
    Relevant Evidence." *Personality and Social Psychology Bulletin* 5
    (2001): 242–273. Accessed March 1, 2012. carlsonschool.umn
    .edu/Assets/71520.pdf.

Bering, Jesse. "My Lust: A Valentine's Day Confession and the Psychology
    of Infatuation." *Scientific American*, February 14, 2011. Accessed
    March 1, 2012. blogs.scientificamerican.com/bering-in-mind
    /2011/02/14/my-lust-a-valentines-day-confession-and-the
    -psychology-of-infatuation/.

Blackburn, Simon. "In Defence of Lust." *New Statesman*, December 15,
    2003.

Diamond, Lisa M. "Emerging Perspectives on Distinctions between
    Romantic Love and Sexual Desire." *Current Directions in
    Psychological Science* 13 (2004): 116–119. www.psych.utah.edu
    /people/people/diamond/Publications/Emerging%20
    Perspectives.pdf.

DiscoveryHealth.com Writers. "Is It Love or Lust?" HowStuffWorks
    .com. Accessed March 1, 2012. health.howstuffworks.com
    /relationships/love/is-it-love-or-lust.htm.

Fisher, Helen E. "Lust, Attraction, and Attachment in Mammalian Reproduction." *Human Nature* 9 (1997): 23–52. Accessed March 1, 2012. www.helenfisher.com/downloads/articles/10lustattraction.pdf.

Fisher, Helen E. et al. "Defining the Brain Systems of Lust, Romantic Attraction, and Attachment." *Archives of Sexual Behavior* 31 (October 2002): 413–419. www.helenfisher.com/downloads/articles/14defining.pdf.

Forster, Jens, Amina Ozelsel, and Kai Epstude. "How Love and Lust Change People's Perceptions of Relationship Partners." *Journal of Experimental Social Psychology* 46 (March 2010): 237–246. Accessed March 1, 2012. www.sciencedirect.com/science/article/pii/S0022103109002121.

Highfield, Roger. "Scientists Locate Brain's 'Censor.'" *The Telegraph*, November 1, 2001. Accessed March 1, 2012. www.telegraph.co.uk/science/science-news/4767055/Scientists-locate-brains-censor.html.

Jacobs, Tom. "Love, But Not Lust, Inspires Creativity." *Pacific Standard*, August 31, 2009. Accessed March 1, 2012. www.psmag.com/culture-society/love-but-not-lust-inspires-creativity-3493/.

Karama, S., et al. "Areas of Brain Activation in Males and Females during Viewing of Erotic Film Excerpts." *Human Brain Mapping* 16 (May 2002): 1–13. Accessed March 1, 2012. www.ncbi.nlm.nih.gov/pubmed/11870922.

NPR. "True Confessions: Men and Women Sin Differently." February 20, 2009. Accessed March 1, 2012. www.npr.org/templates/story/story.php?storyId=100906920.

Platek, Steven M., Julian Paul Keenan, and Todd K. Shackelford. *Evolutionary Cognitive Neuroscience.* Cambridge, MA: MIT Press, 2007.

♥ ♥ ♥ ♥ ♥

⋛ The Dope on Love Drugs: How Aphrodisiacs Work

Dannelke, Lenora. "Take Two Anchos and Call Me in the Morning." Cosmic Chile. Accessed June 29, 2005. www.cosmicchile .com/xdpy/kb/chile-pepper-health-benefits.html.

Hirsch, A.R. and Kim, J.J.: "Effects of Odor on Penile Blood Flow-A Possible Impotence Treatment." *Psychosomatic Medicine*, 57 (1995): 83.

Kamhi, Ellen, PhD, RN. "Natural Aphrodisiacs." NaturalNurse. December 2004. Accessed June 29, 2005. naturalnurse.com /2011/natural-aphrodisiacs.

MotherNature.com. "Inhibited Sexual Desire in Women." Accessed June 29, 2005. www.mothernature.com/archive/centers/detail .cfm?id=2665&term=Sex.

Newman, Judith. "Passion Pills." *Discover*, September 1999.

Ratsch, Christian. *Plants of Love: The History of Aphrodisiacs and a Guide to Their Identification and Use*. Berkeley, CA: Ten Speed Press, 1997.

Short, Stacy. "The Scent of Seduction: How to Use Scent to Your Advantage in the Game of Love." Smell & Taste Treatment and Research Foundation.

Warner, Jennifer. "Aphrodisiacs Make Better Flirts and Lovers." WebMd. February 2006. Accessed June 29, 2005. avacadell.com /internet-press/32-about-dr-ava/press-center/internet/550 -webmd-eat-your-way-to-a-spicier-sex-life.

World Wildlife Fund. "Species: Rhino." Accessed June 29, 2005. worldwildlife.org/species/rhino.

♥ ♥ ♥ ♥ ♥

⋛ The Big O: What Happens in the Brain during an Orgasm?

Brown, Sylvester. "For Women, Sex is No Trip to Disneyland, Dutch Research Says." *St. Louis Post-Dispatch*, November 16, 2003.

Geddes, Donald Porter. *An Analysis of the Kinsey Reports on Sexual Behavior in the Human Male and Female*. New York: New American Library, 1954.

Holstege, Gert, et al. "Brain Activation during Human Male Ejaculation." *Journal of Neuroscience* 23 (October 8, 2003): 9185–9193. www.jneurosci.org/cgi/content/full/23/27/9185.

Komisaruk, Barry R., et al. *The Science of Orgasm*. Baltimore: The Johns Hopkins University Press, 2006.

Le Page, Michael. "Women's Orgasms Are a Turn-Off for the Brain." *New Scientist*, July 1, 2005.

Marett-Carter, Sara. "Sexual Healing." *Flare*, February 2005.

Monroe, Valerie. "Sex is Sublime." *O: the Oprah Magazine*, October 2004.

Nuzzo, Regina. "Science of the Orgasm." *Los Angeles Times*, February 11, 2008. www.latimes.com/features/health/la-he-orgasm11feb11,0,4954575,full.story.

"Orgasm." *Merriam-Webster Dictionary*. 2008. Accessed October 7, 2008. www.merriam-webster.com/dictionary/orgasm?show=0&t=1403634638.

"Orgasm." *Oxford English Dictionary*. Oxford: Oxford University Press, June 2008.

Travis, John. "There's No Faking It." *Science News*, November 29, 2003.

♥ ♥ ♥ ♥ ♥

Is Love at First Sight Possible?

Alexander, Brian. "The Science of Love." MSNBC. February 14, 2006. Accessed April 18, 2011. www.msnbc.msn.com/id/11102123/ns/health-sexual_health/.

Ben-Zeev, Aaron. "Love at First Sight (and First Chat)." *Psychology Today*, May 24, 2008. Accessed April 18, 2011. www.psychologytoday.com/blog/in-the-name-love/200805/love-first-sight-and-first-chat.

Bryner, Jeanna. "People Fall in Love, Brain and Soul." LiveScience. October 26, 2010. Accessed April 18, 2011. www.livescience .com/8821-people-fall-love-brain-soul.html.

Canning, Andrea. "The Science Behind Falling in Love." ABC News. January 17, 2008. Accessed April 18, 2008. abcnews.go.com /GMA/OnCall/story?id=4147929&page=1.

Diamond, Lisa. "Love and Sexual Desire." *Current Directions in Psychological Science* 13 (June 2004): 116-119. cdp.sagepub.com /content/13/3/116.abstract.

The Economist. "The Scent of a Woman (and a Man)." January 10, 2008. Accessed April 18, 2011. www.economist.com/node /10493120?story_id=10493120.

Federation of American Societies for Experimental Biology. "Is Love At First Sight Real? Geneticists Offer Tantalizing Clues." ScienceDaily. April 8, 2009. Accessed April 18, 2011. www .sciencedaily.com/releases/2009/04/090407145203.htm.

Fisher, Helen. "The Realities of Love at First Sight." *O, The Oprah Magazine*, November 2009. Accessed April 18, 2011. www .oprah.com/relationships/Love-at-First-Sight-Helen-Fisher -Love-Column.

Fisher, Helen. "The Science Behind Love at First Sight." Match.com. Accessed April 18, 2011. www.match.com/y/article.aspx?article id=9830&TrackingID=526103&BannerID=696333.

Fisher, Helen E. "The Biology of Attraction." *Psychology Today*, April 1, 1993. Accessed April 18, 2011. www.psychologytoday.com /node/20966.

Melnick, Meredith. "Debunking the Headlines: Falling in Love in 0.2 Sec.? We Don't Think So." *Time*, October 27, 2010. Accessed April 18, 2011/. healthland.time.com/2010/10/27/debunking-the -headlines-falling-in-love-in-0-2-sec-we-dont-think-so/print/.

Moskowitz, Clara. "Love at First Sight Might Be Genetic." LiveScience. April 8, 2009. Accessed April 18, 2011. www.livescience .com/3468-love-sight-genetic.html.

Orr, Deborah. "The Mysterious Power of Attraction." *The Independent*, September 13, 2008. Accessed April 18, 2011. www.independent .co.uk/life-style/love-sex/attraction/the-mysterious-power-of -attraction-926687.html.

Randerson, James. "Love at First Sight Just Sex and Ego, Study Says." *The Guardian*, November 6, 2007. Accessed April 18, 2011. www .guardian.co.uk/science/2007/nov/07/1.

Reuters. "Love at First Sight, or in Half a Second." September 18, 2007. Accessed April 18, 2011. www.reuters.com/article/2007/09/18 /us-beauty-attraction-idUSN1844443620070918.

Rose, Damon. "Love at No Sight." BBC News. May 27, 2009. Accessed April 18, 2011. news.bbc.co.uk/2/hi/uk_news /magazine/8069993.stm.

Rowett Research Institute. "Love at First Sight of Your Body Fat." ScienceDaily. August 13, 2007. Accessed April 18, 2011. www .sciencedaily.com/releases/2007/08/070812095324.htm.

Singleton, Dave. "Love at First Sight: Possible?" Chemistry.com. Accessed April 18, 2011. www.chemistry.com/datingadvice /Love-At-First-Sight.

Syracuse University. "Falling in Love Only Takes about a Fifth of a Second, Research Reveals." ScienceDaily. October 25, 2010. Accessed April 18, 2011. www.sciencedaily.com /releases/2010/10/101022184957.htm.

♥ ♥ ♥ ♥ ♥

Head over Heels: How Love Works

BrainConnection.com. "Gender Related Cultural Confusion—Part 1." April 12, 2004. Accessed February 12, 2005. brainconnection .brainhq.com/gender-related-cultural-confusion-part-1/.

Brennan, Peter. "Something in the Air." The Naked Scientists. February 1, 2006. Accessed February 21, 2013. www.thenakedscientists .com/HTML/articles/article/peterbrennancolumn2.htm/.

Cherry, Kendra. "The Reader Questions About the Psychology of Love." About.com. Accessed February 21, 2013. psychology.about .com/od/loveandattraction/a/love_questions.htm.

Discovery Fit & Health. "Sexual Health." Accessed February 21, 2013. health.howstuffworks.com/sexual-health.

The Economist. "The Science of Love: I Get A Kick Out of You." February 12, 2004. www.economist.com/node/2424049.

Fisher, Helen E. *Why We Love: The Nature and Chemistry of Romantic Love*. New York: Henry Holt, 2004.

Fuentes, Agustin. "What Is Love?" *Psychology Today*, August 14, 2012. www.psychologytoday.com/collections/201208/top-25-list -august-2012/what-is-love.

Garcia, Carlos Yela. "Temporal Course of the Basic Components of Love throughout Relationships." *Psychology in Spain* 2 (1998): 78–86. Accessed February 21, 2013. www.psychologyinspain.com /content/full/1998/9frame.htm.

Landis, Dan, and William A. O'Shea, III. "Cross-Cultural Aspects of Passionate Love." *Journal of Cross-Cultural Psychology* 31 (November 2000): 752–777. Accessed February 21, 2013. jcc .sagepub.com/content/31/6/752.abstract.

McLoughlin, Claire. "Science of Love – Cupid's Chemistry." The Naked Scientists. February 14, 2006. Accessed February 21, 2013. www.thenakedscientists.com/HTML/articles/article/claire mcloughlincolumn1.htm/.

McManamy, John. "The Brain in Love and Lust." McMan's Depression and Bipolar Web. Accessed February 21, 2013. www .mcmanweb.com/love_lust.html.

Oxytocin.org. "Hormone Involved in Reproduction May Have Role in the Maintenance of Relationships." July 14, 1999. Accessed February 21, 2013. www.oxytocin.org/oxytoc.

PBS. "The Science of Love." Accessed February 21, 2013. www.pbs.org /kqed/springboard/segments/59/.

Smuts, Barbara. "The Brain in Love." *Scientific American*, March 22, 2004.

Accessed February 21, 2013. www.scientificamerican.com
/article.cfm?id=the-brain-in-love.

♥ ♥ ♥ ♥ ♥

Love Potion No. 9: Top Five Love Chemicals in the Brain

BBC. "Science of Love." Accessed January 7, 2012. www.bbc.co.uk
/science/hottopics/love/.

Clarke, Terrence. "Pablo Neruda's Love Sonnet 53: A Translation." Red
Room. June 1, 2008. Accessed January 7, 2012. redroom.com
/member/terence-clarke/blog/pablo-nerudas-love-sonnet-53
-a-translation.

Discovery Channel. "Science of Sex Appeal: Dopamine versus
Testosterone." February 4, 2009. Accessed January 7, 2012.
dsc.discovery.com/videos/science-of-sex-appeal-testosterone
-vs-dopamine.html.

Durfee, Rachel. "Sorry, You're Just Not My (Testosterone's) Type."
Popular Science, September 16, 2008. Accessed January 7, 2012.
www.popsci.com/rachel-durfee/article/2008-09/sorry-youre
-just-not-my-testosterones-type.

Fisher, Helen E. et al. "Reward, Addiction, and Emotion Regulation
Systems Associated With Rejection in Love." *Journal of
Neurophysiology* 104 (July 2010): 51–60. Accessed January 7,
2012. jn.physiology.org/content/104/1/51.

Kluger, Jeffrey. "The Science of Romance: Why We Love." *Time*,
January 17, 2008. Accessed January 7, 2012. content.time.com
/time/magazine/article/0,9171,1704672,00.html.

Lite, Jordan. "This Is Your Brain on Love: Lasting Romance Makes an
Impression—Literally." *Scientific American*, January 6, 2009.
Accessed January 6, 2009. www.scientificamerican.com/blog
/post/this-is-your-brain-on-love-lasting-2009-01-06/?id=this
-is-your-brain-on-love-lasting-2009-01-06.

Marazziti, Donatella. "The Neurobiology of Love." *Current Psychiatry Reviews* 1 (November 2005): 331–335.

Obringer, Lee Ann. "How Love Works." HowStuffWorks.com. Accessed January 6, 2012. people.howstuffworks.com/love6.htm.

Yong, Ed. "Of Voles and Men: Exploring the Genetics of Commitment." *Discover Magazine,* September 2, 2008. Accessed January 6, 2012. blogs.discovermagazine.com/notrocketscience/2008/09/02 /of-voles-and-men-exploring-the-genetics-of-commitment/.

♥ ♥ ♥ ♥ ♥

Wooing on the Web: How Online Dating Works

Discovery Fit & Health. "Sexual Health." Accessed February 21, 2013. health.howstuffworks.com/sexual-health.

Dotinga, Randy. "Online Dating Sites Aren't Holding People's Hearts." *The Christian Science Monitor,* January 27, 2005. Accessed February 21, 2013. www.csmonitor.com/2005/0127/p11s02-stin.html.

The Economist. "The Science of Love: I Get A Kick Out of You." February 12, 2004. Accessed February 21, 2013. www.economist.com /node/2424049.

eHarmony. "#1 Trusted Online Dating Site for Singles." Accessed February 21, 2013. www.eharmony.com/.

Fisher, Helen E. *Why We Love: The Nature and Chemistry of Romantic Love.* New York: Henry Holt, 2004.

Fuentes, Agustin. "What Is Love?" *Psychology Today,* August 14, 2012. www.psychologytoday.com/collections/201208/top-25-list -august-2012/what-is-love.

Garcia, Carlos Yela. "Temporal Course of the Basic Components of Love throughout Relationships." *Psychology in Spain* 2 (1998): 78–86. Accessed February 21, 2013. www.psychologyinspain.com /content/full/1998/9frame.htm.

iVillage UK. "Dating Profile Writing Tips." Accessed February 21, 2013. www.ivillage.co.uk/dating-profile-writing-tips/82222.

Landis, Dan, and William A. O'Shea, III. "Cross-Cultural Aspects of Passionate Love." *Journal of Cross-Cultural Psychology* 31 (November 2000): 752–777. Accessed February 21, 2013. jcc .sagepub.com/content/31/6/752.abstract.

McLoughlin, Claire. "Science of Love – Cupid's Chemistry." The Naked Scientists. February 14, 2006. Accessed February 21, 2013. www.thenakedscientists.com/HTML/articles/article/claire mcloughlincolumn1.htm/.

McManamy, John. "The Brain in Love and Lust." McMan's Depression and Bipolar Web. Accessed February 21, 2013. www.mcman web.com/love_lust.html.

Oxytocin.org. "Hormone Involved in Reproduction May Have Role in the Maintenance of Relationships." July 14, 1999. Accessed February 21, 2013. www.oxytocin.org/oxytoc.

PBS. "The Science of Love." Accessed February 21, 2013. www.pbs.org /kqed/springboard/segments/59/.

PR Web. "Is Personality Test a Reliable Tool for Online Dating?" Compatti.com. January 1, 2005. www.prweb.com/releases /2005/01/prweb193081.htm.

Smuts, Barbara. "The Brain in Love." *Scientific American*, March 22, 2004. Accessed February 21, 2013. www.scientificamerican.com /article.cfm?id=the-brain-in-love.

♥ ♥ ♥ ♥ ♥

Agents of Amour: How Matchmakers Work

Abdulrahim, Raja. "His Matches Have Sparks of Tradition." *Los Angeles Times*, September 23, 2011. Accessed August 25, 2012. articles.latimes.com/2011/sep/23/local/la-me-muslim-match makers-20110923.

Berkvist, Robert. "Jerry Brock, 'Fiddler on the Roof' Composer, Dies at 81." *New York Times*, November 3, 2010. Accessed August 25, 2012. www.nytimes.com/2010/11/04/theater/04bock.html?_r=2.

DePaul, Amy and Amy Williams. "The Rise of Arranged Marriage in America." AlterNet. August 8, 2008. Accessed August 25, 2012. www.alternet.org/story/92561/the_rise_of_arranged_marriage_in_america.

Dijkstra, Pieternel and Dick P.H. Barelds. "Do People Know What They Want: A Similar or Complementary Partner?" *Evolutionary Psychology* 6 (2008): 595–602. Accessed August 25, 2012. www.epjournal.net/wp-content/uploads/EP06595602.pdf.

Flanigan, Santana. "Arranged Marriages in India." Post-Colonial Studies @ Emory. Emory University, Fall 2000. Accessed August 25, 2012. www.english.emory.edu/Bahri/Arr.html.

Froelich, Paula. "More Rich, High-Powered Women Are Turning to Matchmakers to Find Love." The Daily Beast. August 15, 2012. Accessed August 25, 2012. www.thedailybeast.com/articles/2012/08/15/more-rich-high-powered-women-are-turning-to-matchmakers-to-find-love.html.

Garone, Elizabeth. "A Matchmaker with a Rich Niche." *The Wall Street Journal*, May 24, 2010. Accessed August 25, 2012. online.wsj.com/article/SB10001424052748703559004575256760383984280.html.

Gerstel, Judy. "Inside the World of Matchmaking." *The Toronto Star*, February 28, 2008. Accessed August 25, 2012. www.thestar.com/life/2008/02/28/inside_the_world_of_matchmaking.html.

Haag, Pamela. "Signs of the Post-Romantic Times: Arranged Marriage Chic?" The Big Think. September 24, 2011. Accessed August 25, 2012. bigthink.com/marriage-30/signs-of-the-post-romantic-times-arranged-marriage-chic?page=all.

Herbert, Wray. "Likes Long Walks in the Woods on Autumn Days." Huffington Post. February 14, 2012. Accessed August 25, 2012. www.huffingtonpost.com/wray-herbert/online-dating_b_1268511.html.

Hicken, Melanie. "Inside the Glamorous Life of a New York City

Matchmaker." *Business Insider*, April 21, 2012. Accessed August 25, 2012. articles.businessinsider.com/2012-04-21/news/313 72135_1_manhattan-s-matchmaker-lisa-ronis-matchmaking -industry.

Indeed. "Matchmaker Salary in New York, NY." Accessed August 25, 2012. www.indeed.com/salary/q-Matchmaker-l-New-York, -NY.html.

Jain, Anita. "Is Arranged Marriage Really Any Worse Than Craigslist?" *New York Times Magazine,* May 21, 2005. Accessed August 25, 2012. nymag.com/nymetro/news/culture/features/11621/.

Joffe-Walt, Chana. "What's a Yenta?" NPR. June 13, 2012. Accessed August 25, 2012. www.npr.org/blogs/money /2012/06/14/154955498/whats-a-yenta/.

Kadden, Barbara Binder and Bruce Kadden. *Teaching Jewish Life Cycle: Traditions and Activities.* Springfield, NJ: Behrman House, 1990.

"Matchmaker." *New World Encyclopedia.* Updated March 19, 2009. Accessed August 25, 2012. www.newworldencyclopedia.org /entry/Matchmaker.

The Matchmaking Institute. "The Matchmaking Industry." Accessed August 25, 2012. www.matchmakingpro.com/dating-match making-industry/.

Rosenfeld, Michael J. "Meeting Online: The Rise of the Internet as a Social Intermediary." Stanford University. April 2010. Accessed August 25, 2012. www.stanford.edu/~mrosenfe/Rosenfeld _How_Couples_Meet_PAA_updated.pdf.

Sherwood, Harriet. "The Jewish Matchmaker." *The Guardian*, January 6, 2011. Accessed August 25, 2012. www.guardian.co.uk/life andstyle/2011/jan/07/jewish-matchmaker-arranged-marriage.

Spindel, Janis and Joan Raymond. "Prowling the Skies in Search of Ringless Fingers." *New York Times*, January 29, 2008. Accessed August 25, 2012. www.nytimes.com/2008/01/29 /business/29flier.html.

Thernstrom, Melanie. "The New Arranged Marriage." *New York Times*

Magazine, February 13, 2005. Accessed August 25, 2012. www
.nytimes.com/2005/02/13/magazine/13MATCHMAKING
.html?pagewanted=all&position.

Toledo, Myrna. "First Comes Marriage, Then Comes Love." ABC News
20/20. January 30, 2009. Accessed August 25, 2012. abcnews
.go.com/2020/story?id=6762309&page=1#.UDpbsqDfWSp.

Van Grove, Jennifer. "Online Dating Is Bigger Than Porn." Mashable.
March 24, 2010. Accessed August 25, 2012. mashable
.com/2010/03/24/online-dating-infographic/.

♥ ♥ ♥ ♥ ♥

Rapid Romance: How Speed Dating Works

8minuteDating.com. "8 Minute Dating Success Stories." Accessed August
23, 2007. www.8minutedating.com/successstories.cgi.

ABC News *20/20*. "Are Pheromones a Secret Weapon for Dating?"
December 9, 2005. abcnews.go.com/2020/Health/story?id=1386825.

BBC News. "Proof Love at First Sight Exists." September 10, 2004.
news.bbc.co.uk/2/hi/health/3643822.stm.

Buzzle.com. "Eight Quick Speed Dating Tips." June 27, 2006. www
.buzzle.com/articles/speed-dating-tips.html.

Cole, Alison. "A Look at Jewish Dating Traditions." EZineArticles.
September 21, 2005. Accessed August 23, 2007. ezinearticles
.com/?A-Look-at-Jewish-Dating-Traditions&id=73889.

Daily, Lisa. "Speed Dating Tips: How to Beat the Love Buzzer." Love
Chat. Accessed August 23, 2007. www.lchat.bravehost.com
/SPEEDDATING.html.

Discovery Fit and Health. "Speed Dating: A New Form of Matchmaking."
Accessed August 23, 2007. health.howstuffworks.com/relationships
/dating/speed-dating-a-new-form-of-matchmaking.htm.

DitchOrDate.com. "How Does Speed Dating Work?" November 16,
2005. www.ditchordate.com/speed-dating/53/.

Einhorn, Rosie and Sherry Zimmerman. "In the Ballpark: When

to Say 'Yes' to a Shidduch Suggestion." SawYouAtSinai. Accessed August 23, 2007. www.sawyouatsinai.com/jewish -matchmaker-article-8.aspx.

Gregorie, Jill. "Love in the Fast Lane." *Generation*. Accessed August 23, 2007. www.subboard.com/generation/articles/116043424551964.asp.

Leadbetter, Ron. "Eros." *Encyclopedia Mythica*. March 3, 1997. Last modified March 2, 2006. www.pantheon.org/articles/e/eros.html.

Love-Tactics.com. "Plan for a Successful Evening with These Speed Dating Tips."Accessed August 23, 2007. www.love-tactics .com/dating/speed-dating-tips.html.

Mythography. "Eros in Greek Mythology." August 2, 2007. www.loggia .com/myth/eros.html.

Pre-Dating.com. "Pre-Dating Success Stories: It Can Happen to You!" Accessed August 23, 2007. www.pre-dating.com/success.php3.

Stanford Graduate School of Business. "The Four-Minute Search for the Perfect Mate." July 1, 2005. www.gsb.stanford.edu/news /research/mktg_simonson_date.shtml.

University of Pennsylvania. "Just in Time for Valentine's Day: Falling in Love in Three Minutes or Less." Penn News. February 11, 2005. www.upenn.edu/pennnews/article.php?id=747.

WorldofJo. "My Speed Dating Experience." February 1, 2007. worldofjo .blogspot.com/2007/02/my-speed-dating-experience.html.

♥ ♥ ♥ ♥ ♥

Ménage à Trois...or More: How Polyamory Works

Anapol, Deborah. "Love Without Limits Blog." *Psychology Today*, August 23, 2010. Accessed February 28, 2011. www.psychologytoday .com/blog/love-without-limits.

Bennett, Jessica. "Only You. And You. And You." *Newsweek*, July 29, 2009. Accessed February 28, 2011. www.newsweek .com/2009/07/28/only-you-and-you-and-you.html.

Cloud, John. "Henry & Mary & Janet & …" *Time*, November 15, 1999.

Accessed February 28, 2011. www.time.com/time/magazine
/article/0,9171,992556,00.html.

Doheny, Kathleen. "The Truth about Open Marriage." WebMD.
November 20, 2007. Accessed February 28, 2011. www.webmd
.com/sex-relationships/features/the-truth-about-open-marriage.

Echlin, Helena. "When Two Just Won't Do." *The Guardian*, November
13, 2003. Accessed February 28, 2011. www.guardian.co.uk
/world/2003/nov/14/gender.uk.

Emens, Elizabeth F. "Monogamy's Law: Compulsory Monogamy and
Polyamorous Existence." *New York University Review of Law &
Social Change* 29 (2004): 277.

Fox, Rose. "Poly 101." Polyamory Online. March 20, 2006. Accessed
February 28, 2011. www.polyamoryonline.org/poly101.html.

Gerard, Jim. "Three's Company: So Is Four or Five." Salon. July 17, 1999.
Accessed February 28, 2011. www.salon.com/1999/07/17
/polyamory/.

Hayes, Veronica. "Sex without Rules." Salon. January 23, 2002. Accessed
February 28, 2011. www.salon.com/2002/01/23/poly/.

Hesse, Monica. "Pairs with Spares." *Washington Post*, February 13, 2008.
Accessed February 28, 2011. www.washingtonpost.com/wp
-dyn/content/article/2008/02/12/AR2008021203072.html.

Lady Chimmerly. "Stranger in a Super-Friendly Land." Salon.
July 17, 1999. Accessed February 28, 2011. www.salon
.com/1999/07/17/conference/.

Langley, Liz. "Whole Lotta Love." Salon. June 14, 2007. Accessed February
28, 2011. www.salon.com/2007/06/14/polyamory_3/.

Leith, William. "Welcome to the World of Polyamory." *The Observer*,
July 9, 2006. Accessed February 28, 2011. www.guardian
.co.uk/lifeandstyle/2006/jul/09/familyandrelationships2.

Miller, Sandra A. "Love's New Frontier." *Boston Globe*, January 3, 2010.
Accessed February 28, 2011. www.boston.com/bostonglobe
/magazine/articles/2010/01/03/loves_new_frontier/.

Newitz, Annalee. "Love Unlimited." *New Scientist*, July 7, 2006. www

.newscientist.com/article/mg19125591.800-love-unlimited -the-polyamorists.html.

PolyFamilies.com. Accessed February 28, 2011. polyfamilies.com.

Weitzman, Geri. "Therapy with Clients Who Are Bisexual and Polyamorous." *Journal of Bisexuality* 6 (2006): 137–164. Accessed February 28, 2011. www.tandfonline.com/doi/abs/10.1300 /J159v06n01_08#preview.

Williams, Alex. "Hopelessly Devoted to You, You and You." *New York Times*, October 5, 2008. Accessed February 28, 2011. www .nytimes.com/2008/10/05/fashion/05polyamory.html.

♥ ♥ ♥ ♥ ♥

Happily Ever After: How Marriage Works

Acevedo, Bianca P. and Arthur Aron. "Does a Long-Term Relationship Kill Romantic Love?" *Review of General Psychology* 13 (2009): 59–65. Accessed January 19, 2012. www.apa.org/pubs /journals/releases/gpr13159.pdf.

Acevedo, Bianca P. et al. "Neural Correlates of Long-Term Intense Romantic Love." *Social Cognitive and Affective Neuroscience* 7 (2012): 145–159. Accessed January 19, 2012. www.helenfisher .com/downloads/articles/Acevedo-et-alLong-term.pdf.

Acs, Gregory and Sandi Nelson. "What Do 'I Do's' Do? Potential Benefits of Marriage for Cohabiting Couples with Children." Urban Institute. May 24, 2004. www.urban.org/publications/311001.html.

Associated Press. "New Jersey Governor Signs Civil Unions into Law." MSNBC. December 21, 2006. www.msnbc.msn.com/id /16309688/.

California Secretary of State. "Domestic Partners Registry." Accessed April 16, 2007. www.sos.ca.gov/dpregistry/.

Clark, Josh. "Is Marriage a Good Investment?" HowStuffWorks.com. Accessed January 19, 2012. money.howstuffworks.com /personal-finance/financial-planning/marriage-investment.htm.

Compton, Todd M. "The Four Major Periods of Mormon Polygamy." The Signature Books Library. Accessed April 16, 2007. signaturebookslibrary.org/?p=425.

District of Columbia Department of Health. "Domestic Partnership FAQ." Accessed April 16, 2007. doh.dc.gov/service/domestic -partnership.

Fisher, Helen. "The Drive to Love." *The New Psychology of Love*. New Haven, CT: Yale University Press, 2006.

Greenfieldboyce, Nell. "Marriage Woes? Husband's Genes May Be at Fault." NPR. September 2, 2008. Accessed January 19, 2012. www.npr.org/templates/story/story.php?storyId=94199631.

Hawaii State Department of Health. "About Marriage Licenses." Accessed April 16, 2007. www.hawaii.gov/health/vital-records/vital -records/marriage/index.html.

Hawaii State Department of Health. "About Reciprocal Beneficiary Relationships." Accessed April 16, 2007. www.hawaii.gov /health/vital-records/vital-records/reciprocal/index.html.

Larson, Aaron. "Marriage Law." ExpertLaw.com. August 2003. www .expertlaw.com/library/family_law/marriage_law.html.

Legal Information Institute. "Marriage." Cornell Law School. August 19, 2010. www.law.cornell.edu/wex/index.php/Marriage.

Legal Information Institute. "Marriage Laws of the Fifty States, District of Columbia and Puerto Rico." Cornell Law School. www.law .cornell.edu/topics/Table_Marriage.htm.

Luscombe, Belinda. "The Five Secrets of Happily Married Parents." *Time*, December 12, 2011. Accessed January 19, 2012. healthland.time .com/2011/12/12/the-five-secrets-of-happily-married-parents/.

Maine Center for Disease Control and Prevention. "Instructions and Information for Declaration of Domestic Partnerships." Accessed April 16, 2007. www.maine.gov/dhhs/mecdc /public-health-systems/data-research/vital-records/documents /pdf-files/dompartinst.pdf.

Marriage Equality USA. "Get the Facts on Marriage." www
.marriageequality.org/facts_at_a_glance.

Miller, Marshall and Dorian Solot. "Common Law Marriage Fact Sheet."
Unmarried Equality. August 2006. www.unmarried.org
/common-law-marriage-fact-sheet/.

Murray, Shailagh. "Gay Marriage Amendment Fails in Senate." *The
Washington Post*, June 8, 2006.

National Conference of State Legislatures. "Common Law Marriage."
Accessed April 16, 2007. www.ncsl.org/research/human
-services/common-law-marriage.aspx.

The National Marriage Project. "Social Indicators of Marital Health
& Well-Being." Accessed January 19, 2012. www.state
ofourunions.org/2011/social_indicators.php#marriage.

New York City Marriage Bureau. "Domestic Partnership Registration."
Accessed April 16, 2007. www.cityclerk.nyc.gov/html
/marriage/domestic_partnership_reg.shtml.

Nolo.com. "Marriage Rights and Benefits." Accessed April 16, 2007.
www.nolo.com/legal-encyclopedia/marriage-rights-benefits
-30190.html.

Parker-Pope, Tara. "The Generous Marriage." *New York Times*,
December 8, 2011. Accessed January 19, 2012. well.blogs
.nytimes.com/2011/12/08/is-generosity-better-than-sex/.

Parker-Pope, Tara. "The Happy Marriage Is the 'Me' Marriage." *New York
Times*. December 31, 2010. Accessed January 19, 2012. www
.nytimes.com/2011/01/02/weekinreview/02parkerpope.html.

Parker-Pope, Tara. "The Science of a Happy Marriage." *New York Times*,
May 10, 2010. Accessed January 19, 2012. well.blogs.nytimes
.com/2010/05/10/tracking-the-science-of-commitment/.

The Salt Lake Tribune. "Living the Principle: Inside Polygamy." Accessed
April 16, 2007.

The Salt Lake Tribune. "Polygamy." (section of articles on the topic).
Accessed April 16, 2007. www.sltrib.com/polygamy.

Shulman, Shlomo. "Guide to the Jewish Wedding." Aish HaTorah. Accessed April 16, 2007. www.aish.com/jl/l/48969841.html.

Slater, Lauren. "True Love." *National Geographic*, February 2006. Accessed January 19, 2012. science.nationalgeographic.com/science /health-and-human-body/human-body/true-love.html.

U.S. Census Bureau. "Marriage and Divorce Data." January 31, 2007. www.census.gov/population/www/socdemo/marr-div.html.

U.S. Census Bureau. "Number, Timing, and Duration of Marriages and Divorces: 2001." Household Economic Studies. February 2005. www.census.gov/prod/2005pubs/p70-97.pdf.

Vermont Secretary of State. "Civil Marriage." Accessed April 16, 2007. www.sec.state.vt.us/municipal/civil_mar.htm.

Words@Random. "The Mavens' Word of the Day: Jumping the Broom." Random House. February 26, 2001. archive.today/pI60v.

WorldWeddingTraditions.com. "Wedding Traditions and Customs around the World." Accessed April 16, 2007. www.world weddingtraditions.com/.

Zimmer, Carl. "Romance Is an Illusion." *Time*, January 17, 2008. Accessed January 19, 2012. content.time.com/time/magazine/ article/0,9171,1704665,00.html

♥ ♥ ♥ ♥ ♥

Betrothed through the Centuries: A Timeline of Marriage

Coontz, Stephanie. *Marriage, a History: How Love Conquered Marriage*. New York: Viking Press, 2005.

Lambda Archives. "History of Marriage Timeline." Accessed February 14, 2011. www.lambdaarchives.us/timelines/marriage/index.htm.

Magnus Hirschfield Archive for Sexology. "History of Marriage in Western Civilization." Accessed February 14, 2011. www2.hu -berlin.de/sexology/ATLAS_EN/html/history_of_marriage _in_western.html.

Offen, Karen. "A Brief History of Marriage: Marriage Laws and Women's Financial Independence." International Museum of Women. Accessed February 14, 2011. www.imow.org/economica/stories/viewStory?storyId=3650.

U.S. Constitution Online. "Constitutional Topic: Marriage." Accessed February 14, 2011. www.usconstitution.net/consttop_marr.html.

♥ ♥ ♥ ♥ ♥

Marriage Multiplied: How Polygamy Works

A&E. *A&E Investigative Reports: Inside Polyga,* 1999. Directed by Enos, Lisa. Nicholas Claxton Productions.

Gray, Patrick J., ed. "Ethnographic Atlas Codebook." *World Cultures* 1 (1998): 86–136.

Ross, Rick. "A Brief History of the Polygamists in Colorado City, Arizona and Hildale, Utah." Cult Education Institute. April 5, 2002. culteducation.com/group/1099-polygamist-groups/16741-a-brief-history-of-the-polygamists-in-colorado-city-arizona-and-hildale-utah.html.

Schultz, Emily A. and Robert H. Lavenda. *Cultural Anthropology*. Oxford, UK: Oxford University Press, 2004.

Tanner, Sandra and Jerald Tanner. "Covering Up Mormon Polygamy." *Salt Lake City Messenger*, August 1998.

Tracy, Kathleen. *The Secret Story of Polygamy*. Naperville, IL: Sourcebooks, 2001.

♥ ♥ ♥ ♥ ♥

Achy Breaky Heart: How Breakups Work

Baxter, Leslie A. "Gender Differences in the Heterosexual Relationship Rules Embedded in Break-Up Accounts." *Journal of Social and Personal Relationships* 3 (September 1986): 289–306. Accessed January 26, 2012. psycnet.apa.org/psycinfo/2004-20749-003.

Brinig, Margaret F. and Douglas W. Allen. "'These Boots Are Made for Walking': Why Most Divorce Filers Are Women." *American Law and Economics Review* 2 (2000): 126–169. Accessed January 26, 2012. www.unc.edu/courses/2010fall/econ/586/001/Readings /Brinig.pdf.

Byron, Lee. "Breakups—The Visual Miscellaneum." Lee Byron blog. November 2009. Accessed January 26, 2012. leebyron.com /what/breakups/.

Choo, Patricia, Timothy Levine and Elaine Hatfield. "Gender, Love Schemas, and Reactions to Romantic Breakups." *Handbook of Gender Research*, special edition of *Journal of Social Behavior and Personality* 11 (1996): 143–160. Accessed January 26, 2012. www.elainehatfield.com/96.pdf.

Cosier, Susan. "Breakups Cloud Sense of Self, Study Finds." LiveScience. March 12, 2010. Accessed January 27, 2012. www.livescience .com/6211-breakups-cloud-sense-study-finds.html.

Crook, Brittani. "A Roller Coaster of Love: Examining Perceptions of Intimacy, Commitment, and Satisfaction in On-Again, Off-Again Relationships." Texas State University-San Marcos. May 2010. Accessed February 8, 2012. digital.library.txstate.edu /handle/10877/3940.

Dailey, Rene M. et al. "On-Again/Off-Again Dating Relationships: How Are They Different from Other Dating Relationships?" *Personal Relationships* 16 (March 2009): 23–47. Accessed January 26, 2012. onlinelibrary.wiley.com/doi/10.1111/j.1475-6811 .2009.01208.x/abstract.

Dailey, Rene M. et al. "A Qualitative Analysis of On-Again/Off-Again Romantic Relationships: It's Up and Down, All Around." *Journal of Social and Personal Relationships* 26 (June 2009): 443–466. Accessed January 26, 2012. spr.sagepub.com /content/26/4/443.abstract.

Davis, Deborah, Phillip R. Shaver, and Michael L. Vernon. "Physical, Emotional, and Behavioral Reactions to Breaking Up: The

Roles of Gender, Age, Emotional Involvement, and Attachment Style." *Personality and Social Psychology Bulletin* 29 (July 2003): 871-84. Accessed January 26, 2012.

Durex. "Sexual Wellbeing Global Survey." Accessed January 26, 2012. www. durex.com/en-lv/sexualwellbeingsurvey/pages/default.aspx.

Fisher, Helen. *Lost Love: The Nature of Romantic Rejection*. New Brunswick, NJ: Rutgers University Press, 2006.

Fisher, Helen E. et al. "Reward, Addiction, and Emotional Regulation Systems Associated with Rejection in Love." *Journal of Neurophysiology* 104 (May 5, 2010), 51–60. Accessed January 26, 2012. www.helenfisher.com/downloads/articles/Fisher-et -al-Rejection.pdf.

Haupt, Angela. "Health Buzz: Romantic Rejection Causes Physical Pain." *U.S. News & World Report*, March 29, 2011. Accessed January 26, 2012. health.usnews.com/health-news/family-health/pain /articles/2011/03/29/health-buzz-romantic-rejection-causes -physical-pain.

Kross, Ethan et al. "Social Rejection Shares Somatosensory Representations with Physical Pain." *Proceedings of the National Academy of Sciences* 108 (April 12, 2011): 6270–6275. Accessed January 26, 2012. www.pnas.org/content/108/15/6270.full.

Lannutti, Pamela J. and Kenzie A. Cameron. "Beyond the Breakup: Heterosexual and Homosexual Post-Dissolutional Relationships." *Communications Quarterly* 50 (Spring 2002): 153–170. Accessed January 27, 2012.

Nashawaty, Chris. "Elizabeth Taylor and 'The Scandal of the Century': A Look Back at the Romance That Rocked the World." *Entertainment Weekly*, March 23, 2011. Accessed January 26, 2012. popwatch.ew.com/2011/03/23/elizabeth-taylor -richard-burton-scandal/.

Northwestern University. "Breaking Up May Not Be as Hard as the Song Says." ScienceDaily. August 20, 2007. Accessed January 27, 2012. www.sciencedaily.com/releases/2007/08/070820123608.htm.

Spielman, S.S., G. Macdonald, and A.E. Wilson. "On the Rebound: Focusing on Someone Helps New Anxiously Attached Individuals Let Go of a Loss." *Personal Sociology and Psychology Bulletin* 35 (October 2009): 1382–1394. Accessed February 2, 2012. www.ncbi.nlm.nih.gov/pubmed/19625631.

Svoboda, Elizabeth. "A Thoroughly Modern Guide to Breaking Up." *Psychology Today*, January 1, 2011. Accessed January 26, 2012. www.psychologytoday.com/articles/201012/the-thoroughly -modern-guide-breakups?page=4.

Tashiro, Ty and Patricia Frazier. "I'll Never Be in a Relationship Like That Again: Personal Growth Following Romantic Relationship Breakups." *Personal Relationships* 10 (March 2003): 113–128. Accessed January 26, 2012.

♥ ♥ ♥ ♥ ♥

Can You Die of a Broken Heart?

Derrick, Dawn. "The 'Broken Heart Syndrome': Understanding Takotsubo Cardiomyopathy." *Critical Care Nurse* 29 (February 2009): 49-57.

Dote, K. et al. "Myocardial Stunning Due to Simultaneous Multivessel Coronary Spasms: A Review of 5 Cases." *Journal of Cardiology* 21 (1991): 203-14.

Grady, Denise. "Sudden Stress Breaks Heart, a Report Says." *New York Times*, October 11, 2011. Accessed January 6, 2012. www.nytimes.com/2005/02/10/health/10heart.html?page wanted=print&position.

Grimes, Jessica Collins. "Study Offers a Closer Look At 'Broken Heart Syndrome.'" *Medical News Today*, March 28, 2009. Accessed January 6, 2012. www.eurekalert.org/pub_releases/2009-03/l -mab032409.php.

Johns Hopkins Medicine. "'Broken Heart' Syndrome: Real, Potentially Deadly But Recover Quickly." February 9, 2005. Accessed

January 6, 2012. www.hopkinsmedicine.org/press_releases /2005/02_10_05.html.

Martikainen, Pekka and Tampani Valkonen. "Mortality after Death of Spouse in Relation to Duration of Bereavement in Finland." *Journal of Epidemiology and Community Health* 50 (1996): 264–268. Accessed January 6, 2012. www.ncbi.nlm.nih.gov/pmc /articles/PMC1060281/pdf/jepicomh00183-0036.pdf.

Mayo Clinic Staff. "Broken Heart Syndrome." Mayo Clinic. February 10, 2011. Accessed January 6, 2012. www.mayoclinic.com/health /broken-heart-syndrome/DS01135/METHOD=print.

Metcalf, Eric. "Can You Die of a Broken Heart?" WebMD. Accessed January 6, 2012. www.webmd.com/heart/features/broken -heart-syndrome-stress-cardiomyopathy.

National Heart, Lung and Blood Institute. "What Is Cardiac Catheterization?" May 1, 2006. Accessed January 26, 2012. www.nhlbi.nih.gov/health/health-topics/topics/cath/.

Parker-Pope, Tara. "Health and the Broken Heart." *New York Times*, June 1, 2010. Accessed January 6, 2012. well.blogs.nytimes .com/2010/06/01/health-and-the-broken-heart/.

Salamon, Maureen. "Women More Prone to 'Broken Heart' Syndrome: Study." HealthDay. November 16, 2011. Accessed January 6, 2012. consumer.healthday.com/senior-citizen-information-31 /age-health-news-7/women-more-prone-to-broken-heart -syndrome-study-658921.html.

Winslow, Ron. "Hearts Can Actually Break." *The Wall Street Journal*, February 9, 2010. Accessed January 6, 2012. online.wsj.com/article/SB10001424 052748703615904575053443911673752.html.

♥ ♥ ♥ ♥ ♥

⋗ Why Do Breakup Songs Hurt So Good?

Binns-Turner, Pamela. "Perioperative Music and Its Effects on Anxiety, Hemodynamics, and Pain in Women Undergoing

Mastectomy." *AANA Journal* 79 (August 2011): S21–S26. Accessed January 23, 2012. www.aana.com/newsandjournal /Documents/perioperative_08res11_pS21-S27.pdf.

Bradshaw, D.H. et al. "Individual Differences in the Effect of Music Engagement in Responses to Painful Stimulation." *Pain* 12 (December 2011): 1262–1273. Accessed January 23, 2012. www.ncbi.nlm.nih.gov/pubmed/22071366.

Fisher, Helen et al. "Reward, Addiction, and Emotion Regulation Systems Associated with Rejection in Love." *Journal of Neurophysiology* 104 (May 1, 2010): 51–60. Accessed January 23, 2012. www .helenfisher.com/downloads/articles/Fisher-et-al-Rejection.pdf.

Limjoco, Victor. "Music for Pain." *Discover*, August 8, 2006. Accessed January 23, 2012. discovermagazine.com/2006/aug/musicnopain.

Nilsson, S. et al. "School-Aged Children's Experiences of Postoperative Music Medicine on Pain, Distress, and Anxiety." *Pediatric Anesthesia*. 19 (December 2009): 1184–1190. January 23, 2012. www.ncbi.nlm.nih.gov/pubmed/19863741.

O'Connor, Anahad. "Really? The Claim: Listening to Music Can Relieve Pain." *New York Times*, January 2, 2012. Accessed January 23, 2012. well.blogs.nytimes.com/2012/01/02/really-the-claim -listening-to-music-can-relieve-pain/.

Roy, Mathieu, Isabelle Peretz, and Pierre Rainville. "Emotional Valence Contributes to Music-Induced Analgesia." *Pain* 134 (January 2008): 140–147. Accessed January 23, 2012. www.sciencedirect .com/science/article/pii/S0304395907001856.

Totten, John. "The Comfort Food of Pretzel Logic: Regulating Emotions with Steely Dan." The Other Journal, October 24, 2011. Accessed January 23, 2012. theotherjournal.com/2011/10/24/the-comfort-food-of-pretzel-logic-regulating-emotion-with-steely-dan/.

Wilcox, Christie. "Time—and Brain Chemistry—Heals All Wounds." *Scientific American*, October 24, 2011. Accessed January 23, 2012. blogs.scientificamerican.com/science-sushi/2011/10/24 /brain_chemistry_emotional_wounds/.

♥ ♥ ♥ ♥ ♥

⅀ Top Five Love Lies

Ackerman, Joshua M., Vladas Griskevicius, and Norman P. Li. "Let's Get Serious: Communication Commitment in Romantic Relationships." *Journal of Personal and Social Psychology* 100 (2011): 1079–1094. Accessed February 2, 2012. www.csom .umn.edu/assets/165677.pdf.

Bittman, Kate. "Take Cover: A New Set of Rules Is Coming." *The New Yorker*, June 24, 2011. Accessed February 2, 2012. www .newyorker.com/online/blogs/books/2011/06/take-cover-a -new-set-of-rules-is-coming.html.

Brady, Lois Smith. "Ellen Fein and Lance Houpt." *New York Times*, August 9, 2008. Accessed February 2, 2012. www .nytimes.com/2008/08/10/fashion/weddings/10VOWS .html?pagewanted=all.

Eastwick, Paul W. et al. "Mispredicting Distress Following Romantic Breakup: Revealing the Time Course of the Affective Forecasting Error." *Journal of Experimental Psychology*. 44 (June 29, 2007). Accessed February 2, 2012. faculty.wcas.northwestern.edu/eli -finkel/documents/ForecastingPageProofs8-14-07.pdf.

Fisher, Helen. *Why Him? Why Her? Finding Real Love by Understanding Your Personality Type*. New York: Macmillan, 2009.

Frith, Maxine. "Women Are Happiest with First Love and Men with 'Serial Monogamy,' Study Finds." *The Independent*, December 22, 2003. Accessed February 2, 2012. www.independent.co.uk/ news/uk/this-britain/women-are-happiest-with-first-love-and- men-with-serial-monogamy-study-finds-577451.html.

Galperin, Andrew and Martie Haselton. "Predictors of How Often and When People Fall in Love." *Evolutionary Psychology* 8 (2010): 5–28. Accessed February 2, 2012. www.epjournal.net/wp -content/uploads/ep080528.pdf.

Harrison, M.A. and J.C. Shortall. "Women and Men in Love: Who

Really Feels It and Says It First?" *Journal of Social Psychology* 151 (Nov–Dec 2011): 727–736. Accessed February 2, 2012. www.ncbi.nlm.nih.gov/pubmed/22208110.

Hill, Amelia. "Why We Can Never Recover from First Love." *The Guardian*, January 17, 2009. Accessed February 2, 2012. www.guardian.co.uk/lifeandstyle/2009/jan/18/relationships-love.

Law, Sally. "The Truth on Whether 'Opposites Attract.'" LiveScience. March 27, 2009. Accessed February 2, 2012. www.livescience.com/5351-truth-opposites-attract.html.

Lehrer, Jonah. "Opposites Don't Attract (And That's Bad News)." *Wired*, January 16, 2012. Accessed February 2, 2012. www.wired.com/wiredscience/2012/01/opposites-dont-attract-and-thats-bad-news/.

Luscombe, Belinda. "The Cruelest Study: Why Breakups Hurt." *Time*, July 9, 2010. Accessed February 2, 2012. content.time.com/time/health/article/0,8599,2002688,00.html.

Norton, Cherry. "Old Flames Reunited Make the Most Lasting Marriages." *The Independent*, August 14, 2000. Accessed February 2, 2012. www.independent.co.uk/news/science/old-flames-reunited-make-the-most-lasting-marriages-711097.html.

Oppenheimer, Mark. "Married, With Infidelities." *New York Times Magazine*, June 30, 2011. Accessed February 2, 2012. www.nytimes.com/2011/07/03/magazine/infidelity-will-keep-us-together.html?pagewanted=all.

Romance Writers of America. "Romance Literature Statistics: Industry Statistics." 2010. Accessed February 2, 2012. www.rwa.org/p/cm/ld/fid=580.

Than, Ker. "Wild Sex: Where Monogamy Is Rare." LiveScience. November 2, 2006. Accessed February 2, 2012. www.livescience.com/1135-wild-sex-monogamy-rare.html.

The Rules. "Top 10 Rules." TheRulesBook. Accessed February 2, 2012.

Scalise, Kathleen. "After the Breakup, Your 'First Love' Never Really Leaves You, according to Student Research at UC Berkeley." UC Berkeley Campus News. February 7, 2001.

Accessed February 2, 2012. berkeley.edu/news/media/releases/2001/02/07_love.html.

Tresniowski, Alex. "There Goes the Bride." *People*, April 16, 2001. Accessed February 2, 2012. www.people.com/people/archive/article/0,,20134135,00.html.

Viegas, Jennifer. "Monogamous Animals Often Have Unattractive Partners." Discovery News. February 1, 2011. Accessed February 2, 2012. news.discovery.com/animals/monogamous-animals-mating-attractiveness-110201.html.

♥ ♥ ♥ ♥ ♥

Liar, Liar: Does Cheating Run in the Family?

CBS News. "Nightmare in Camelot." February 11, 2009. Accessed February 21, 2012. www.cbsnews.com/stories/2003/07/02/politics/main561365.shtml.

Cherkas, Lynn F. et al. "Genetic Influences on Female Infidelity and Number of Sexual Partners in Humans: A Linkage and Association Study of the Role of the Vasopressin Receptor Gene (AVPR1A)." *Twin Research* 7 (August 30, 2004): 649-658.

Garcia, Justin R, et al. "Associations between Dopamine D4 Receptor Gene Variation with Both Infidelity and Sexual Promiscuity." *PLoS ONE* 5 (November 30, 2010). Accessed February 21, 2012. www.plosone.org/article/info:doi/10.1371/journal.pone.0014162.

Havlíček, Jan et al. "Correlates of Extra-Dyadic Sex in Czech Heterosexual Couples: Does Sexual Behavior of Parents Matter?" *Archives of Sexual Behavior* 40 (December 2011): 1153–1163. Accessed February 21, 2012. www.ncbi.nlm.nih.gov/pubmed/22033668.

Karolinska Institute. "Infidelity Gene? Genetic Link To Relationship Difficulties Found." ScienceDaily. September 2, 2008. Accessed February 21, 2012. www.sciencedaily.com/releases/2008/09/080902161213.htm.

Korda, Michael. "Joe Kennedy's Hollywood Fling." The Daily Beast. February 4, 2009. Accessed February 21, 2012. www.the dailybeast.com/articles/2009/02/04/joe-kennedys-hollywood -fling.html.

McGreal, Chris. "Author Alleges Jackie and Bobby Kennedy Began Affair after JFK Assassination." *The Guardian*, July 7, 2009. Accessed February 21, 2012. www.guardian.co.uk/world/2009/jul/07 /bobby-jackie-kennedy-jfk-book.

Peele, Stanton. "The Top Seven Kennedy Sex Scandals." *Psychology Today*, May 21, 2008. Accessed February 21, 2012. www .psychologytoday.com/blog/addiction-in-society/200805 /reckless-sex-and-power-iii-the-top-seven-kennedy-sex-scandals.

Reuters. "Former Intern reveals 18-Month Affair with JFK." February 6, 2012. Accessed February 21, 2012. www.reuters.com /article/2012/02/06/us-jfk-affair-idUSTRE8151VS20120206.

Time. "Chappaquiddick: Suspicions Renewed." May 11, 1970. Accessed February 21, 2012. content.time.com/time/magazine/article /0,9171,878212,00.html.

Tsapelas, Irene, Helen E. Fisher, and Arthur Aron. "Infidelity: When, Where, Why." In *The Dark Side of Close Relationships II* by William Cupach and Brian Spitzberg. London: Routledge, August 24, 2010.

Vedantam, Shankar. "Study Links Gene Variant in Men to Marital Discord." *The Washington Post*, September 2, 2008. Accessed February 21, 2012. www.washingtonpost.com /wp-dyn/content/article/2008/09/01/AR2008090102087 .html?nav=hcmodule.

♥ ♥ ♥ ♥ ♥

Conclusion: What's the Future of Virtual Sex?

Born, Matt. "Serves You Right, Duped MPs Told." *The Telegraph*, December 21, 2001. Accessed February 5, 2011. www

.telegraph.co.uk/news/uknews/1365930/Serves-you-right
-duped-MPs-told.html.

Central Intelligence Agency (CIA). "World." CIA World Factbook, 2010.
Accessed February 6, 2011. www.cia.gov/library/publications/
the-world-factbook/.

CuteCircuit. "Hug Shirt," 2012. Accessed January 27, 2012. cutecircuit
.com/collections/the-hug-shirt/.

Halverson, Nic. "Kiss Transmitter Lets You Make Out over the Internet."
Discovery News. May 5, 2011. Accessed January 27, 2012.
news.discovery.com/tech/kiss-transmitter-lets-you-make-out
-internet-110505.html.

Ku, Jeonghun et al. "A Data Glove with Tactile Feedback for FMRI
of Virtual Reality Experiments." *Cyberpsychology & Behavior* 5
(October 2003): 497-508. Accessed January 27, 2012. www
.ncbi.nlm.nih.gov/pubmed/14583125.

Lynn, Regina. "Ins and Outs of Teledildonics." *Wired*, September 24,
2004. Accessed January 27, 2012. archive.wired.com/culture
/lifestyle/commentary/sexdrive/2004/09/65064.

Madrigal, Alexis. "Researchers Want to Add Touch, Taste and Smell to
Virtual Reality." *Wired*, March 4, 2009. Accessed January 27,
2012. www.wired.com/wiredscience/2009/03/realvirtuality/.

Nuzzo, Regina. "Call Him Doctor 'Orgasmatron.'" *Los Angeles Times*,
February 11, 2010. Accessed January 27, 2012. www.latimes
.com/la-he-orside11feb11,1,5212599.story.

Ruvolo, Julie. "The Internet Is For Porn (So Let's Talk About It)." *Forbes*,
May 20, 2011. Accessed January 27, 2012. www.forbes.com
/sites/julieruvolo/2011/05/20/the-internet-is-for-porn-so
-lets-talk-about-it/.

Shell, Barry. "Wire Up Your Sense of Smell: How the Internet Is Changing
the World of Perfumery." *Scientific American*, September 9,
2011. Accessed January 27, 2012. blogs.scientificamerican
.com/guest-blog/2011/09/09/wire-up-your-sense-of-smell
-how-the-internet-is-changing-the-world-of-perfumery/.

Stephens, Walter. *Demon Lovers: Witchcraft, Sex, and the Crisis of Belief*. Chicago: University of Chicago Press, 2003.

Terdiman, Daniel. "Kinect Sex Is Here, Game Company Says." CNET. December 15, 2010. Accessed January 27, 2012. news.cnet .com/8301-13772_3-20025804-52.html.

Wilson, Daniel H. "How Haptics Will Change the Way We Interact With Machines." *Popular Mechanics*, October 1, 2009. Accessed January 27, 2012. www.popularmechanics.com/technology /gadgets/news/4253368.

ABOUT
HOWSTUFFWORKS

Your walk, your smell, or even your nicely symmetrical face can attract potential suitors faster than you can say "sex appeal." But do you have any idea why? In response to pleas from our love-struck and love-thirsty readers, the curious minds at HowStuffWorks.com dove into the steamy science of attraction and love. The subjects are more complicated than you might think. After all, humans have been doing this love and marriage thing for ages, way before breakup songs and Facebook status changes entered the picture, but we still don't have all the answers. This volume tackles everything from aphrodisiacs to polyamory to broken hearts. Read this book and you'll know what to do the next time someone sets your heart racing.

HowStuffWorks.com is an award-winning digital source of credible, unbiased, and easy-to-understand explanations of how the world actually works. Founded in 1998, the site is now an online resource for millions of people of all ages. From car engines to search engines, from cell phones to stem cells, and thousands of subjects in between,

HowStuffWorks.com has it covered. In addition to comprehensive articles, our helpful graphics and informative videos walk you through every topic clearly and objectively. Our premise is simple: demystify the world and do it in a clear-cut way that anyone can understand.

If you enjoyed *The Real Science of Sex Appeal: Why We Love, Lust, and Long for Each Other*, check out the rest of the series from Sourcebooks and HowStuffWorks.com!

STUFF YOU MISSED IN HISTORY CLASS: A GUIDE TO HISTORY'S BIGGEST MYTHS, MYSTERIES, AND MARVELS

DISCOVER HISTORY AS YOU NEVER KNEW IT!

For years, the hosts of *Stuff You Missed in History Class*, the popular podcast from the award-winning website HowStuffWorks, have been giving listeners front-row seats to some of the most astonishing and amazing stories the human record has to offer. Now, tens of millions of downloads later, they present the ultimate crash course in world history. Featuring the best of the podcast, this engaging book explores the coolest and craziest scandals, myths, lies, and crimes your history teachers never wanted you to know. Discover:

- ♥ How medieval torture devices really worked
- ♥ Whether the CIA tested LSD on unsuspecting Americans
- ♥ How ninja work—the real Assassin's Creed
- ♥ Which culture invented both pasta and hang gliders (Hint: It's not the Italians!)
- ♥ And more!

Come along for the adventure and stay for the education.
You haven't met history like this before!

FUTURE TECH, RIGHT NOW: X-RAY VISION, MIND CONTROL, AND OTHER AMAZING STUFF FROM TOMORROW

FROM X-RAY VISION TO MIND READING, THE FUTURE IS COMING ON FAST!

Flying cars! Teleporting! Robot servants! Wouldn't you love any of these? You're in luck because they may be closer to reality than you think. Based on the best of HowStuffWorks' popular podcasts *TechStuff*, *Stuff from the Future*, and *Stuff to Blow Your Mind*, this dynamic book reveals the science of our future, from mind control and drugs that can make you smarter to textbooks that talk to you and even robotic teammates. Discover:

- How telekinesis and digital immortality work
- Whether computers could replace doctors one day
- What robot servants and coworkers will look like
- Five of the coolest future car technologies
- What we will do for fun in 2050
- And much more!

Come explore the coolest and craziest technology of the future.

THE SCIENCE OF SUPERHEROES AND SPACE WARRIORS: LIGHTSABERS, BATMOBILES, KRYPTONITE, AND MORE!

DO YOU HAVE WHAT IT TAKES TO BE A SUPERHERO?

You picked out your superpower years ago. You can change into your costume in seconds. You could take out a Sith Lord with your lightning-quick lightsaber moves. Not so fast! Before you can start vanquishing bad guys, it's important to be schooled in the science of saving the world. Come learn the science behind your favorite superheroes and supervillains and their ultracool devices and weapons—from Batmobiles and warp speed to lightsabers, Death Stars, and kryptonite—and explore other cool technologies from the science fiction realm in this dynamic book. Discover:

- ♥ How Batman and the Batmobile really work
- ♥ 10 *Star Trek* technologies that actually came true
- ♥ Whether warp speed and lightsabers are really possible
- ♥ If Superman would win against Harry Potter, Sith Lords, and even Chuck Norris!
- ♥ How new liquid body armor can make us superhuman
- ♥ And more!

Prepare to do battle with the world's most evil masterminds!